Vibration of Buildings to Wind and Earthquake Loads

T. Balendra

Vibration of Buildings to Wind and Earthquake Loads

With 108 Figures

Springer-Verlag
London Berlin Heidelberg New York
Paris Tokyo Hong Kong
Barcelona Budapest

T. Balendra, BSc(Eng), PhD, CEng
Department of Civil Engineering
National University of Singapore
10 Kent Ridge Crescent
Singapore 0511

ISBN-13: 978-1-4471-2057-5   e-ISBN-13: 978-1-4471-2055-1
DOI: 10.1007/978-1-4471-2055-1

British Library Cataloguing in Publication Data
Balendra, Thambirajah
    Vibration of Buildings to Wind and Earthquake Loads
    I. Title
    690.21

Library of Congress Cataloging-in-Publication Data
Balendra, T. (Thambirajah)
    Vibration of buildings to wind and earthquake loads / T. Balendra
        p. cm.
    Includes bibliographical references and index.
        1. Wind resistant design.   2. Earthquake resistant design.   I. Title.
    TA658.48.B34   1993
    693'.85—dc20   93-15460

© Springer-Verlag London Limited 1993
Softcover reprint of the hardcover 1st edition 1993

The publisher makes no representation, express or implied, with regard to the accuracy
of the information contained in this book and cannot accept any legal responsibility
or liability for any errors or omissions that may be made.

Typeset by Integral Typesetting, Great Yarmouth, Norfolk
Printed by Antony Rowe Ltd, Chippenham, Wiltshire
69/3830-543210 Printed on acid-free paper

To Karuna,
Viknesh and Ganesh

Preface

Recent advances in the development of high-strength materials coupled with more advanced computational methods and design procedures have led to a new generation of tall buildings which are slender and light. These buildings are very sensitive to the two most common dynamic loads – winds and earthquakes. The primary requirement for a successful design is to provide safety during infrequent events such as severe earthquakes. In addition, serviceability requirements such as human comfort and integrity of structural components during strong winds or moderate earthquakes need to be addressed. As such, the design of tall buildings subject to winds and moderate earthquakes is governed by stiffness. To resist strong earthquakes and suffer only repairable damage, the current trend is to incorporate weak elements in the structural system so that they act as structural fuses in the event of extreme loading. Thus the efficiency of the design depends on the type of structural systems chosen for construction. From teaching, research and the consultancy experience of the author, an attempt has been made in this book to provide a well balanced and broad coverage of the information needed for the design of structural systems for wind- and earthquake-resistant buildings. To this end, the basic concepts in structural dynamics and structural systems, assessment of wind and earthquake loads acting on the system, evaluation of the system response to such dynamic loads and finally the design for extreme loading are presented. Numerical examples are included in the text. The book is intended for graduate students of structural engineering as well as for practising structural engineers.

In writing this book, I am indebted to my family who has supported me enthusiastically throughout. I would like to acknowledge my gratitude to all who have helped me in the preparation of this manuscript, in particular Lim Saukoon, Yeo Kathy, Liu Ding Mei and Jayanti Krishnasamy for typing the manuscript and Mei Yin for drafting the figures. My gratitude is extended to my sons, Viknesh and Ganesh, for allowing their father to write with such little interruption.

T. Balendra
Singapore 1993

Contents

Chapter 1

Fundamentals of Structural Dynamics

1.1 Introduction

Buildings, in addition to gravity loads, are liable to be subjected to time-varying loads arising from winds, earthquakes, machinery, road and rail traffic, construction work, explosions, etc. These loads are dominant over certain frequency ranges. For example, Fig. 1.1 depicts the frequency content of turbulent winds and earthquakes. The influence of these loads on buildings depends on the dynamic characteristics of the buildings in relation to the dominant frequency range of the loading. For example, a stiff building with a period of 0.5 s would only be slightly affected by the wind, but the effect of an earthquake on this building could be serious. On the other hand, for a tall and flexible building of period 5 s, moderate earthquakes may have no serious effects, whereas winds could have a significant effect and would control the design. In order to determine the extent to which the building would be affected by time-varying or dynamic loads, the building must be analysed using the

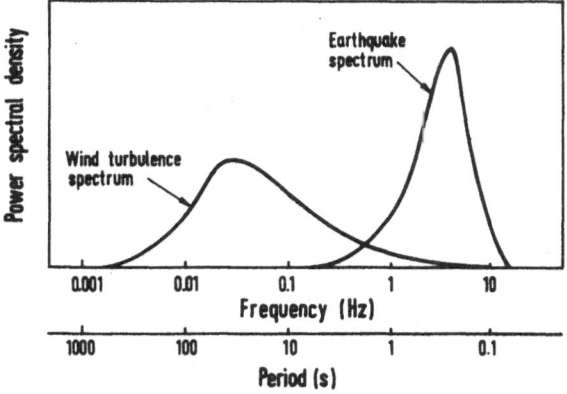

Fig. 1.1. Spectral densities of earthquakes and winds.

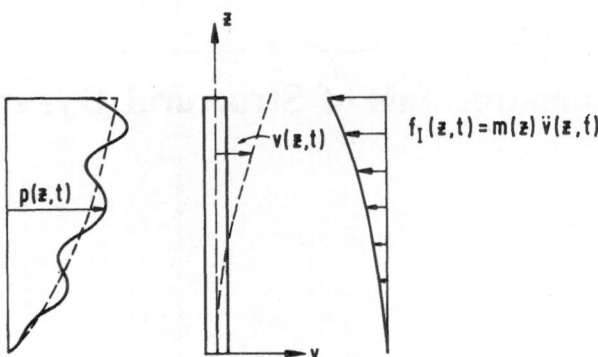

Fig. 1.2. A vertical structure subjected to dynamic loading.

principle of structural dynamics. In a dynamics problem, the applied loadings (and hence structural responses such as deflection, internal forces, stress, etc.) vary with time. Thus, unlike a statics problem, a dynamics problem requires a separate solution at every instant of time. Consider a vertical structure of mass $m(z)$ per unit height subjected to a dynamic load due to wind gust as shown in Fig. 1.2. The applied load $p(z, t)$ produces a time-varying deflection $v(z, t)$ which would involve acceleration $\ddot{v}(z, t)$. These accelerations generate inertia forces $f_i(z, t) = m(z)\ddot{v}(z, t)$ which oppose the motion. Thus the structure may be considered as subjected to two loadings, namely the applied load and the inertia forces. The inertia forces are the essential characteristic of a structural dynamics problem. The magnitude of the inertia forces depends on (a) the rate of loading, (b) the stiffness of the structure and (c) the mass of the structure.

If the loading is applied slowly, the inertia forces are small in relation to the applied loading and may be neglected, and thus the problem can be treated as static. If the loading is rapid, the inertia forces are significant and their effect on the resulting response must be determined by dynamic analysis.

Generally, structural systems are continuous and their physical properties or characteristics are distributed. However, in many instances it is possible to simplify the analysis by replacing the distributed characteristics by discrete characteristics, using the technique of lumping. Thus mathematical models of structural dynamics problems can be divided into two major types:

1. discrete systems with finite degrees of freedom
2. continuous systems with infinite degrees of freedom.

However in the latter case, a good approximation to the exact solution can be obtained using a finite number of appropriate shape functions. The analysis of a discrete system with one degree of freedom will be described first, followed by a multi-degree-of-freedom system. Subsequently the analysis of a continuous system will be presented.

1.2 One-degree-of-freedom System

1.2.1 Equation of Motion

Consider a one-storey building subjected to a lateral load $P(t)$ as shown in Fig. 1.3a. If the floor girder is assumed to be rigid, and axial deformations are neglected, the lateral displacement $v(t)$ of the floor is the only degree of freedom. When the mass m is assumed to be lumped at the floor level, the physical problem can be represented by the mechanical model shown in Fig. 1.3b. In this model, the spring represents the stiffness of the storey and the dashpot represents the viscous damping – the energy-loss mechanisms of the structure. The free body diagram of the mass is shown in Fig. 1.3c where, according to D'Alembert's principle, the inertia force is such as to oppose the motion. Setting the summation of forces to zero yields the equation of motion of the system as

$$m\ddot{v} + c\dot{v} + kv = P(t) \tag{1.1}$$

where m is the mass, c, the viscous damping coefficient and k the lateral stiffness of the building. The consistent units for mass, stiffness, damping coefficient, force and displacement are kilogram, Newton/metre, Newton second/metre, Newton and metre, respectively.

1.2.2 Free Vibration

Vibration in the absence of external forces is called the *free vibration*, which is caused by initial conditions such as initial displacement $v(o)$, initial velocity $\dot{v}(o)$ or a combination of initial displacement and initial velocity. The equation of

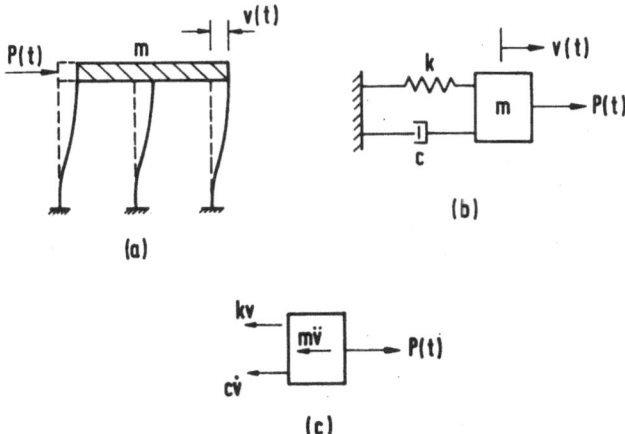

Fig. 1.3. One-storey building subjected to lateral load: (a) physical model, (b) mechanical model and (c) free body diagram.

motion for free vibration takes the form

$$\ddot{v} + \frac{c}{m}\dot{v} + \frac{k}{m}v = 0 \tag{1.2a}$$

or

$$\ddot{v} + 2\zeta\omega\dot{v} + \omega^2 v = 0 \tag{1.2b}$$

where $\omega = \sqrt{k/m}$ is the frequency of undamped vibration in radians/s, and ζ is the damping ratio, expressing the damping as a fraction of critical damping $(2m\omega)$. The critical damping is the smallest damping for which the system returns to its original position without oscillation. The solution to Eq. (1.2) may be expressed as [1.1]

$$v(t) = e^{-\zeta\omega t}\left[\frac{\dot{v}(o) + v(o)\zeta\omega}{\omega_D} \text{Sin } \omega_D t + v(o) \text{Cos } \omega_D t\right] \tag{1.3}$$

where $\omega_D = \omega(1 - \zeta^2)^{1/2}$ is the frequency of the damped vibration. Since the damping in civil engineering structures is small, less than 10% ($\zeta < 0.1$), $\omega_D \simeq \omega$ and thus the period of oscillation T can be taken to be $2\pi/\omega$. The frequency of oscillation f expressed in cycles per second (hertz) is $1/T$.

The free vibration response obtained from Eq. (1.3) is shown in Fig. 1.4. The direction of motion repeats periodically with period T. The magnitude decays exponentially with time owing to the presence of viscous damping.

The ratio between two successive positive or negative peaks can be shown to be

$$\frac{v_n}{v_{n+1}} = e^{\zeta\omega T} \tag{1.4}$$

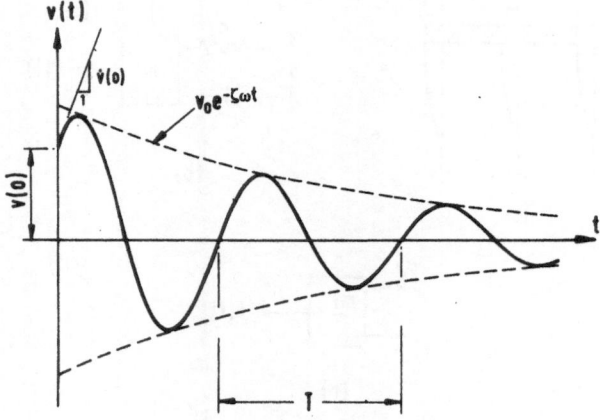

Fig. 1.4. Damped free vibration.

Taking the natural logarithm on both sides of Eq. (1.4) yields

$$\delta = \ln\left(\frac{v_n}{v_{n+1}}\right) = 2\pi\zeta \tag{1.5a}$$

Thus

$$\zeta = \frac{\delta}{2\pi} \tag{1.5b}$$

where δ is called the *logarithmic decrement*. Thus, from Eq. (1.5a), the damping in the system can be estimated. For better accuracy the response peaks which are several cycles apart must be considered, and then the damping ratio is determined from

$$\zeta = \frac{1}{2\pi N}\ln\left(\frac{v_n}{v_{n+N}}\right) \tag{1.6}$$

where N is the number of cycles between the amplitude peaks.

1.2.3 Response to Harmonic Loading

When the external forcing function is of the form

$$P(t) = P_0\, e^{i\omega_f t} \tag{1.7}$$

where P_0 is the amplitude and ω_f is the frequency of the applied loading, the solution to Eq. (1.1) is obtained by setting

$$v(t) = v_0^*\, e^{i\omega_f t} \tag{1.8}$$

where v_0^* is complex. Substituting Eqs. (1.7) and (1.8) into Eq. (1.1) yields

$$v_0^* = H(\omega_f)\frac{P_0}{k} \tag{1.9}$$

where the frequency response function $H(\omega_f)$ is given by

$$H(\omega_f) = \frac{1}{\left\{1 - \left(\dfrac{\omega_f}{\omega}\right)^2\right\} + i2\zeta\left(\dfrac{\omega_f}{\omega}\right)} \tag{1.10}$$

The modulus of the frequency response function is known as the mechanical admittance function. It takes the form

$$|H(\omega_f)| = \frac{1}{\left[\left\{1 - \left(\dfrac{\omega_f}{\omega}\right)^2\right\}^2 + 4\zeta^2\left(\dfrac{\omega_f}{\omega}\right)^2\right]^{1/2}} \tag{1.11}$$

Equation (1.8) may be expressed as

$$v(t) = |v_0^*| \, e^{i(\omega_f t - \theta)} \tag{1.12}$$

where the displacement amplitude and the phase angle are given by

$$|v_0^*| = \frac{P_0}{k} |H(\omega_f)| \tag{1.13a}$$

and

$$\theta = \text{Tan}^{-1} \left\{ \frac{2\zeta\left(\dfrac{\omega_f}{\omega}\right)}{1 - \left(\dfrac{\omega_f}{\omega}\right)^2} \right\} \tag{1.13b}$$

When the applied load is $P_0 \cos \omega_f t$, which is the real part of $P_0 \, e^{i\omega_f t}$, the steady-state solution is obtained by taking the real part of Eq. (1.11), namely

$$v(t) = |v_0^*| \, \text{Cos}(\omega_f t - \theta) \tag{1.14}$$

Similarly if the applied load is $P_0 \sin \omega_f t$, then the corresponding solution is

$$v(t) = |v_0^*| \, \text{Sin}(\omega_f t - \theta) \tag{1.15}$$

The steady-state solution given in Eqs. (1.14) and (1.15) must be added to the transient solution arising from the initial conditions [Eq. (1.3)]. However, the latter will become vanishingly small after a certain period of time, as a result of damping, and thus insignificant in many practical cases.

Since P_0/k is the displacement that would be obtained if the load is applied statically, the ratio $|v_0^*|/(P_0/k)$ is called the *dynamic magnification factor*, D. The plot of dynamic magnification factor against ω_f/ω is shown in Fig. 1.5 for different damping ratios.

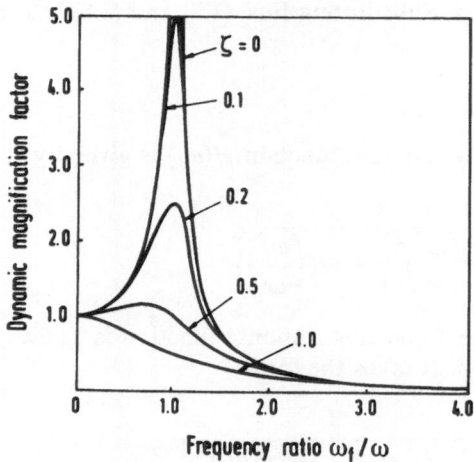

Fig. 1.5. Magnification factor for different damping ratios.

The condition $\omega_f = \omega$ is called *resonance*. At resonance, $D = 1/2\zeta$. As is seen from Fig. 1.5, the response of the undamped system tends to infinity at resonance. With damping, the response is bounded at resonance, however the magnitude is quite large. For instance, at 1% damping ($\zeta = 0.01$) the value of D is 50, and at 2% damping ($\zeta = 0.02$) the corresponding value of D is 25. Thus the resonance condition should be avoided in designs where the structure is subjected to a forcing function with a constant excitation frequency. When a structure is subjected to a complex loading consisting of several frequency components, as in the case of earthquake or wind loading, it is not possible to avoid resonance because one of the exciting frequencies coincides with the structural frequency. However, if the energy associated with such a frequency is small, then it is possible to design the structure to withstand the effects of resonance.

1.2.4 Response to Arbitrary Loading

Earthquake loading and wind loading are irregular and non-periodic. There are various ways to determine the response to such loadings. One method is to assume that the irregular load is made up of several impulses, as shown in Fig. 1.6. By superposing the response due to each impulse, the total response can be obtained.

To determine the response due to a single impulse, consider the Dirac delta function $\delta(t - \tau)$ acting at time $t = \tau$, as shown in Fig. 1.7. The properties of the Dirac delta function are

$$\delta(t - \tau) = 0, \qquad t \neq \tau \tag{1.16a}$$

$$\int_{-\infty}^{\infty} \delta(t - \tau)\, dt = 1 \tag{1.16b}$$

The time interval ε over which the Dirac delta function is not equal to zero is infinitesimally small. When $\varepsilon \to 0$, the magnitude is undefined, however, the

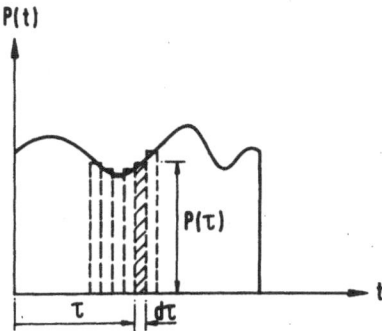

Fig. 1.6. Arbitrary loading discretized into impulses.

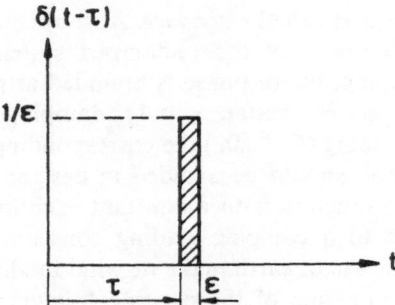

Fig. 1.7. Dirac delta function.

area under the curve is bounded and is equal to 1. The unit of Dirac delta function is 1/seconds.

According to Eqs. (1.16a) and (1.16b), the impulse $\hat{P}$ (in Newton seconds) acting at time $t = \tau$ can be converted to an impulsive force $P(t)$ defined for all time t, as

$$P(t) = \hat{P}\,\delta(t - \tau) \tag{1.17}$$

To determine the response due to an impulse at $t = 0$, we set $\tau = 0$ in the above expression. The governing equation now takes the form

$$m\ddot{v} + c\dot{v} + kv = \hat{P}\,\delta(t) \tag{1.18}$$

Assuming zero initial conditions, namely $v(o) = \dot{v}(o) = 0$, at the time of application of the impulse, integrating Eq. (1.18) over the interval ε and taking the limit of $\varepsilon \to 0$ [1.2]:

$$\lim_{\varepsilon \to 0} \int_0^\varepsilon (m\ddot{v} + c\dot{v} + kv)\,dt = \lim_{\varepsilon \to 0} \int_0^\varepsilon \hat{P}\,\delta(t)\,dt = \hat{P} \tag{1.19}$$

$$\lim_{\varepsilon \to 0} \int_0^\varepsilon m\ddot{v}\,dt = \lim_{\varepsilon \to 0} m[\dot{v}(\varepsilon)] = m\dot{v}(o^+) \tag{1.20a}$$

$$\lim_{\varepsilon \to 0} \int_0^\varepsilon c\dot{v}\,dt = \lim_{\varepsilon \to 0} c[v(\varepsilon)] = cv(o^+) \tag{1.20b}$$

$$\lim_{\varepsilon \to 0} \int_0^\varepsilon kv\,dt = \lim_{\varepsilon \to 0} (kv\varepsilon) = 0 \tag{1.20c}$$

where $\dot{v}(o^+)$ is the instantaneous change in velocity and $v(o^+)$ is the instantaneous change in displacement. As a finite time is required for displacement to develop, $v(o^+) = 0$ for a system starting from rest. Thus Eq. (1.19) yields

$$\dot{v}(o^+) = \frac{\hat{P}}{m} \tag{1.21}$$

The physical interpretation of Eq. (1.21) is that the impulsive force produces an instantaneous change in the velocity. In fact, this is the manner in which the initial velocities are imparted to a system possessing inertia.

Since the external force is absent for $t > 0^+$, the response due to the impulse is the free vibration due to the initial condition given by Eq. (1.21). Using the expression for free vibration given in Eq. (1.3), the response can be expressed as

$$v(t) = \frac{\hat{P}}{m\omega_D} e^{-\zeta\omega t} \operatorname{Sin} \omega_D t, \qquad t > 0$$

$$= 0, \qquad\qquad\qquad t \le 0 \qquad\qquad (1.22)$$

The response due to unit impulse at $t = 0$ is called the *impulsive response* which is obtained by setting $\hat{P} = 1$ in Eq. (1.22) as

$$h(t) = \frac{1}{m\omega_D} e^{-\zeta\omega t} \operatorname{Sin} \omega_D t, \qquad t > 0$$

$$= 0, \qquad\qquad\qquad t \le 0 \qquad\qquad (1.23)$$

Since the response due to unit impulse at $t = \tau$ is $h(t - \tau)$, referring to Fig. 1.6, the response due to an impulse of $P(\tau)\, \Delta\tau$ acting at time $t = \tau$ is given by

$$\Delta v(t, \tau) = P(\tau) h(t - \tau)\, \Delta\tau, \qquad t > \tau \qquad\qquad (1.24)$$

Each impulse in Fig. 1.6 will contribute to the total response. Summing the effects of all the impulses and letting $\Delta\tau \to 0$:

$$v(t) = \int_0^t P(\tau) h(t - \tau)\, d\tau$$

$$= \frac{1}{m\omega_D} \int_0^t P(\tau)\, e^{-\zeta\omega(t-\tau)} \operatorname{Sin} \omega_D(t - \tau)\, d\tau \qquad\qquad (1.25)$$

This is called the *Duhamel integration* or *convolution integral*. If the initial conditions are non-zero at the time of application of the arbitrary loading, then the response is obtained by adding Eq. (1.3) to Eq. (1.25).

Example 1.1. Derive the expressions for displacement time history when the shear building shown in Fig. 1.8 is subjected to a half Sine pulse. Neglect damping and assume zero initial conditions.

Solution. The forcing function is given as

$$P(t) = P_0 \operatorname{Sin} \omega_f t, \qquad 0 \le t \le t_0$$

$$= 0, \qquad\qquad\qquad t > t_0$$

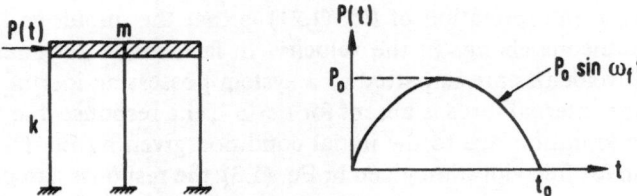

Fig. 1.8. One-storey shear building subjected to half sine pulse.

Thus for $0 \le t \le t_0$ (phase I), Eq. (1.25) yields

$$v(t) = \frac{P_0}{m\omega} \int_0^t \text{Sin } \omega_f \tau \text{ Sin } \omega(t - \tau) \, d\tau$$

$$= \frac{P_0}{k} \frac{1}{(1 - \beta^2)} [\text{Sin } \omega_f t - \beta \text{ Sin } \omega t]$$

where $\beta = \dfrac{\omega_f}{\omega}$

For $t > t_0$ (phase II):

$$v(t) = \frac{P_0}{m\omega} \int_0^{t_0} \text{Sin } \omega_f \tau \text{ Sin } \omega(t - \tau) \, d\tau + \int_{t_0}^t (o) \, d\tau$$

$$= \frac{P_0}{k} \frac{1}{(1 - \beta^2)} [\beta(\text{Cos } \omega_f t_0 - \text{Cos } \omega t_0) \text{ Sin } \omega(t - t_0)$$

$$+ (\text{Sin } \omega_f t_0 - \beta \text{ Sin } \omega t_0) \text{ Cos } \omega(t - t_0)]$$

Note that, during phase II, the system executes free vibration for which the initial conditions are the displacement and velocity at $t = t_0$ (end of phase I). Thus, the response for $t > t_0$ can also be obtained from Eq. (1.3) as

$$v(t) = \frac{\dot{v}(t_0)}{\omega} \text{Sin } \omega(t - t_0) + v(t_0) \text{ Cos } \omega(t - t_0)$$

where

$$v(t_0) = \frac{P_0}{k} \frac{1}{(1 - \beta^2)} [\text{Sin } \omega_f t_0 - \beta \text{ Sin } \omega t_0]$$

$$\dot{v}(t_0) = \frac{P_0}{k} \frac{1}{(1 - \beta^2)} [\omega_f \text{ Cos } \omega_f t_0 - \omega_f \text{ Cos } \omega t_0]$$

1.3 Multi-degree-of-freedom System

1.3.1 Equation of Motion

Consider a three-storey building idealized as a discrete system. If the floor girders are assumed to be rigid and the axial deformation of the columns is neglected, then this system will have three degrees of freedom, one degree of translation per floor, as shown in Fig. 1.9a.

Let the lumped masses at the first, second and third storeys be m_1, m_2 and m_3. If the lateral stiffnesses of the first, second and third storeys are k_1, k_2 and k_3 and the corresponding viscous damping coefficients are c_1, c_2 and c_3, then this physical system can be represented by the mechanical system shown in Fig. 1.9b. Referring to the free body diagram shown in Fig. 1.9c, three equations of motion, one for each mass, can be written. These three equations can be expressed in matrix form as follows:

$$[M]\{\ddot{v}\} + [C]\{\dot{v}\} + [K]\{v\} = \{P\} \tag{1.26}$$

where

$$\{P\} = \begin{Bmatrix} P_1 \\ P_2 \\ P_3 \end{Bmatrix} ; \text{ the force vector}$$

$$\{v\} = \begin{Bmatrix} v_1 \\ v_2 \\ v_3 \end{Bmatrix} ; \text{ the displacement vector}$$

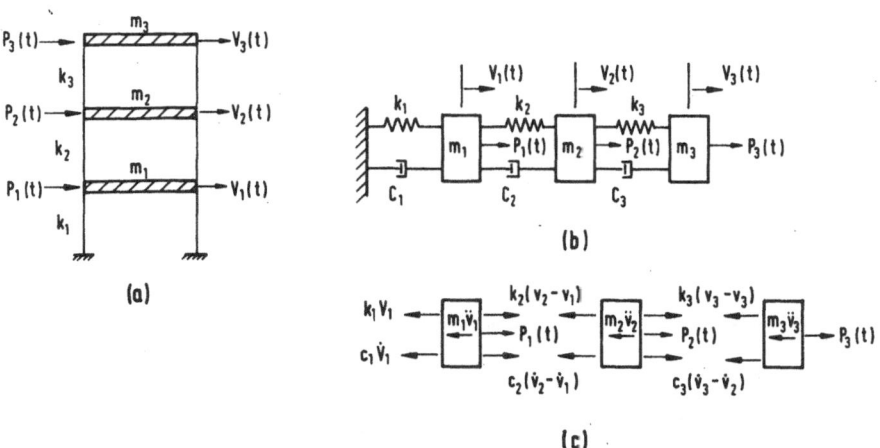

Fig. 1.9. Multistorey shear building subjected to lateral load: (a) physical model, (b) mechanical model and (c) free body diagram.

$$[M] = \begin{bmatrix} m_1 & & \\ & m_2 & \\ & & m_3 \end{bmatrix}; \text{ the mass matrix}$$

$$[C] = \begin{bmatrix} c_1 + c_2 & -c_2 & \\ -c_2 & c_2 + c_3 & -c_3 \\ & -c_3 & c_3 \end{bmatrix}; \text{ the damping matrix}$$

and

$$[K] = \begin{bmatrix} k_1 + k_2 & -k_2 & \\ -k_2 & k_2 + k_3 & -k_3 \\ & -k_3 & k_3 \end{bmatrix}; \text{ the stiffness matrix}$$

In the governing equation, the mass matrix is diagonal because of the lumped mass idealization adopted. The stiffness and damping matrices are symmetrical with positive diagonal terms and with the largest term being in the diagonal.

A convenient method to determine the elements of the stiffness matrix is through the definition of the stiffness influence coefficient k_{ij}, which is defined as the force required at degree of freedom i to produce a unit displacement at degree of freedom j and zero displacements at all the remaining degrees of freedom.

For example, if we set $v_1 = 1$ and $v_2 = v_3 = 0$ to the shear building in Fig. 1.9a, then from the deformation configuration shown in Fig. 1.10a, the first column of the stiffness matrix is determined. Similarly, setting $v_2 = 1$, $v_1 = v_3 = 0$ and $v_3 = 1$, $v_1 = v_2 = 0$ as shown in Figs. 1.10b and 1.10c yields respectively the second and third columns of the stiffness matrix.

In practice, the damping matrix is constructed by assuming it to be proportional to the mass matrix and/or the stiffness matrix. The proportionality constants are determined through damping ratios in different modes of vibrations obtained from vibration tests conducted on similar structures.

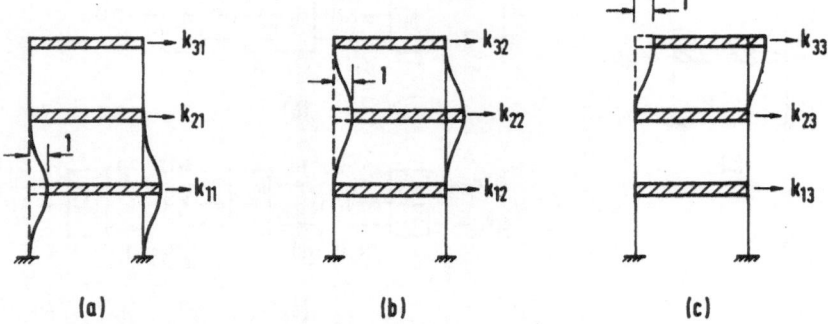

Fig. 1.10. Determination of stiffness matrix: (a) $v_1 = 1$, $v_2 = v_3 = 0$, (b) $v_2 = 1$, $v_1 = v_3 = 0$ and (c) $v_3 = 1$, $v_1 = v_2 = 0$.

For a system with N degrees of freedom, the governing equation of motion, Eq. (1.26), consists of matrices of size $N \times N$ representing N coupled simultaneous second-order differential equations. The solution is obtained through a step-by-step integration procedure [1.1] with certain assumptions about the variation of acceleration, velocity and displacement over the selected time step. However, in the case of the linear system, a more convenient and elegant method to solve this set of equations is through modal analysis, where the solution is obtained as the superposition of the contribution from different modes of vibrations. This method will be elaborated in subsequent sections of this chapter.

1.3.2 Undamped Free Vibration

According to Eq. (1.26), the governing equation for undamped free vibration takes the form

$$[M]\{\ddot{v}\} + [K]\{v\} = \{0\} \tag{1.27}$$

In analogy to the single-degree-of-freedom system, assuming that the free vibration motion of the multi-degree-of-freedom system is simple harmonic, let

$$\{v(t)\} = \{\vartheta\} \, \text{Sin}(\omega t + \theta) \tag{1.28}$$

This implies that we are looking for a solution in which all the coordinates $v_j(t), j = 1, 2, \ldots, N$ execute a synchronous motion, i.e. all the coordinates have the same time dependence. Thus the general configuration of the motion does not change, except for the amplitude.

Substituting Eq. (1.28) into Eq. (1.27) yields

$$[[K] - \omega^2[M]]\{\vartheta\} = \{0\} \tag{1.29}$$

Determining the values of ω^2 for the non-trivial solution of Eq. (1.29) is known as the *eigen value problem*. Non-trivial solution is possible only if the determinant of the matrix vanishes. Thus

$$|[K] - \omega^2[M]| = 0 \tag{1.30}$$

The above equation is called the *characteristic equation*. Expanding this equation leads to a polynomial equation of N-th degree in ω^2. In general, there are N distinct roots or eigen values to this equation. For real symmetric, positive definite mass and stiffness matrices which pertain to stable structural systems, the eigen values are real and positive. If the eigen values arranged in ascending order are denoted as $\omega_1^2, \omega_2^2, \ldots, \omega_N^2$ $(\omega_1^2 > \omega_2^2, \ldots, > \omega_N^2)$, then the square roots of these quantities are the frequencies of the 1st, 2nd, $\ldots$, N-th modes of vibration. The mode having the lowest frequency is called the *fundamental mode*.

Associated with each frequency ω_r there is a non-trivial vector $\{\vartheta_r\}$ satisfying

$$[K]\{\vartheta_r\} - \omega_r^2[M]\{\vartheta_r\} = \{0\}, \qquad r = 1, 2, \ldots, N \tag{1.31}$$

The vectors $\{\vartheta_r\}$, $r = 1$ to N are known as the *eigen vectors*. Equation (1.31) satisfies identically, since the frequency is determined from this equation. Thus it consists of only $N - 1$ independent equations and hence the amplitude of all the elements of vector $\{\vartheta_r\}$ cannot be determined. However the ratios between the elements are determined uniquely, which implies that the shape of $\{\vartheta_r\}$ is known. Thus we could express

$$\{\vartheta_r\} = c_r\{\phi_r\} \tag{1.32}$$

where $\{\phi_r\}$ is the r-th mode shape corresponding to frequency ω_r, and c_r is the unknown amplitude.

According to Eq. (1.28), the response due to the r-th mode of vibration is given by

$$\{v(t)\}_r = c_r\{\phi_r\} \operatorname{Sin}(\omega_r t + \theta_r) \tag{1.33}$$

Since for a linear system, the general solution is the sum of the individual solutions, the general solution for free vibration takes the form

$$\{v(t)\} = \sum_{r=1}^{N} \{v(t)\}_r = \sum_{r=1}^{N} c_r\{\phi_r\} \sin(\omega_r t + \theta_r) \tag{1.34}$$

in which the unknown amplitude c_r, $r = 1$ to N, and phase angle θ_r, $r = 1$ to N, are determined from $2N$ initial conditions, namely $\{v(o)\}$ and $\{\dot{v}(o)\}$.

Example 1.2. Calculate the frequencies and mode shapes of the three-storey shear building in Fig. 1.11. The storey stiffness and the floor masses are as indicated in the figure.

Solution. The mass and stiffness matrices for the building are determined from Eq. (1.26) as

$$[M] = 10^3 \begin{bmatrix} 2.0 & & \\ & 1.5 & \\ & & 1.0 \end{bmatrix} \text{kg}$$

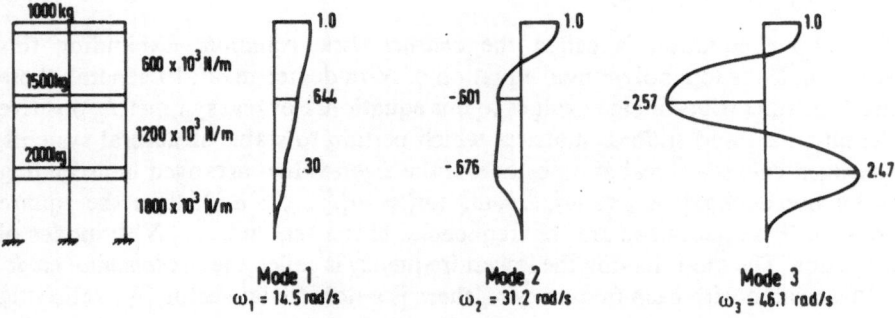

Fig. 1.11. Modal shapes of three-storey shear building.

$$[K] = 10^3 \begin{bmatrix} 3000 & -1200 & 0 \\ -1200 & 1800 & -600 \\ 0 & -600 & 600 \end{bmatrix} \text{N/m}$$

Setting the determinant to zero, as in Eq. 1.30, leads to

$$\lambda^3 - 5.5\lambda^2 + 7.5\lambda - 2 = 0$$

where

$$\lambda = \frac{\omega^2}{600}$$

The roots of the above polynomial equation are $\lambda_1 = 0.351$, $\lambda_2 = 1.61$ and $\lambda_3 = 3.54$. The values correspond to $\omega_1 = 14.5 \, \text{rad/s}$, $\omega_2 = 31.2 \, \text{rad/s}$ and $\omega_3 = 46.1 \, \text{rad/s}$. The corresponding mode shapes are found to be

$$\{\phi_1\} = \begin{Bmatrix} 0.30 \\ 0.644 \\ 1.0 \end{Bmatrix}, \quad \{\phi_2\} = \begin{Bmatrix} -0.676 \\ -0.601 \\ 1.0 \end{Bmatrix}, \quad \{\phi_3\} = \begin{Bmatrix} 2.47 \\ -2.57 \\ 1.0 \end{Bmatrix}$$

Note that mode 1 does not intersect the zero displacement line, whereas mode 2 intersects it once and mode 3 twice.

Orthogonality of Mode Shapes

Consider two distinct solutions ω_r^2, $\{\phi_r\}$ and ω_s^2, $\{\phi_s\}$ of the eigen value problem expressed in Eq. (1.29). Thus we have

$$[K]\{\phi_r\} = \omega_r^2[M]\{\phi_r\} \tag{1.35a}$$

$$[K]\{\phi_s\} = \omega_s^2[M]\{\phi_s\} \tag{1.35b}$$

Premultiplying Eq. (1.35a) by $\{\phi_s\}^T$ and Eq. (1.35b) by $\{\phi_r\}^T$ yields

$$\{\phi_s\}^T[K]\{\phi_r\} = \omega_r^2\{\phi_s\}^T[M]\{\phi_r\} \tag{1.36a}$$

$$\{\phi_r\}^T[K]\{\phi_s\} = \omega_r^2\{\phi_r\}^T[M]\{\phi_s\} \tag{1.36b}$$

Taking the transpose of Eq. (1.36b) leads to

$$\{\phi_s\}^T[K]\{\phi_r\} = \omega_s^2\{\phi_s\}^T[M]\{\phi_r\} \tag{1.36c}$$

Subtracting Eq. (1.36c) from Eq. (1.36a) yields

$$(\omega_r^2 - \omega_s^2)\{\phi_s\}^T[M]\{\phi_r\} = 0 \tag{1.37}$$

Since $\omega_r \neq \omega_s$, Eq. (1.37) implies

$$\{\phi_s\}^T[M]\{\phi_r\} = 0, \quad r \neq s \tag{1.38}$$

Substituting Eq. (1.38) into Eq. (1.36c) leads to

$$\{\phi_s\}^T[K]\{\phi_r\} = 0, \qquad r \neq s \tag{1.39}$$

Equations (1.38) and (1.39) are the orthogonality conditions of the modal vectors.

Response to Initial Condition

The free vibration response of a multi-degree-of-freedom system is given by Eq. (1.34) as

$$\{v(t)\} = \sum_{r=1}^{N} c_r\{\phi_r\} \, \text{Sin}(\omega_r t + \theta_r) \tag{1.40}$$

If $\{v(o)\}$ and $\{\dot{v}(o)\}$ are the initial displacement and velocities respectively, then applying Eq. (1.40), at $t = 0$

$$\{v(o)\} = \sum_{r=1}^{N} c_r\{\phi_r\} \, \text{Sin} \, \theta_r \tag{1.41a}$$

$$\{\dot{v}(o)\} = \sum_{r=1}^{N} c_r\omega_r\{\phi_r\} \, \text{Cos} \, \theta_r \tag{1.41b}$$

Premultiplying Eqs. (1.41) by $\{\phi_r\}^T[M]$ and invoking the orthogonality condition given in Eq. (1.38) lead to

$$c_r \, \text{Sin} \, \theta_r = \frac{\{\phi_r\}^T[M]\{v(o)\}}{\{\phi_r\}^T[M]\{\phi_r\}} \tag{1.42a}$$

$$c_r \, \text{Cos} \, \theta_r = \frac{1}{\omega_r} \frac{\{\phi_r\}^T[M]\{\dot{v}(o)\}}{\{\phi_r\}^T[M]\{\phi_r\}} \tag{1.42b}$$

Applying Eq. (1.42), Eq. (1.40) yields

$$\{v(t)\} = \sum_{r=1}^{N} \left(\frac{1}{\omega_f} \frac{\{\phi_r\}^T[M]\{\dot{v}(o)\}}{\{\phi_r\}^T[M]\{\phi_r\}} \, \text{Sin} \, \omega_r t + \frac{\{\phi_r\}^T[M]\{v(o)\}}{\{\phi_r\}^T[M]\{\phi_r\}} \, \text{Cos} \, \omega_r t \right)\{\phi_r\} \tag{1.43}$$

From the above equation, the free vibration response can be determined for any given initial conditions. For instance, if the initial displacement vector resembles a particular modal vector, say $\{\phi_s\}$, and the initial velocity vector is zero, then setting $\{v(o)\} = \alpha\{\phi_s\}$ and $\{\dot{v}(o)\} = \{0\}$ in Eq. (1.43) yields

$$\{v(t)\} = \alpha\{\phi_s\} \, \text{Cos} \, \omega_s t \tag{1.44}$$

Thus the system executes a harmonic oscillation at natural frequency ω_s and the system configuration resembles the s-th mode all the time. This implies that the natural modes can be excited independently of one another by the appropriate choice of initial conditions.

1.3.3 Forced Vibration Response

Undamped System

For a system with N degrees of freedom, the governing equation of motion, according to Eq. (1.26), takes the form

$$[M]\{\ddot{v}\} + [K]\{v\} = \{P(t)\} \tag{1.45}$$

For a linear system, the solution to the above equation is obtained more conveniently by modal analysis which uncouples the system of differential equations of motion into a set of independent differential equations, each representing a single-degree-of-freedom system. This is achieved by transforming the physical coordinates $\{v(t)\}$ into modal coordinates $\{q(t)\}$, namely

$$\begin{array}{ccc} \{v(t)\} = & [\Phi] & \{q(t)\} \\ N \times 1 & N \times n & n \times 1 \end{array} \tag{1.46}$$

where

$$[\phi] = [\{\phi_1\}\{\phi_2\}\cdots\{\phi_n\}] \tag{1.47}$$

is the mode shape matrix comprising the first n eigen vectors (where n is much smaller than the total number of degrees of freedom N) determined from the corresponding free vibration problem.

Substituting Eq. (1.46) into Eq. (1.45) and premultiplying the resulting equation by $[\phi]^T$ yields

$$[\phi]^T[M][\phi]\{\ddot{q}\} + [\phi]^T[K][\phi]\{q\} = [\phi]^T\{P(t)\} \tag{1.48}$$

According to the orthogonality conditions given in Eqs. (1.38) and (1.39), Eq. (1.48) takes the form:

$$m_r^*\ddot{q}_r + k_r^*q_r = p_r^*(t), \qquad r = 1, 2, \ldots, n \tag{1.49}$$

or

$$\ddot{q}_r + \omega_r^2 q_r = \frac{p_r^*(t)}{m_r^*}, \qquad r = 1, 2, \ldots, n$$

where

$$m_r^* = \{\phi_r\}^T[M]\{\phi_r\} \tag{1.50a}$$

$$k_r^* = \{\phi_r\}^T[K]\{\phi_r\} \tag{1.50b}$$

$$\{p_r^*(t)\} = \{\phi_r\}^T\{P(t)\} \tag{1.50c}$$

Equation (1.49) consists of n uncoupled equations, each representing the motion of a single-degree-of-freedom system in modal coordinates.

According to Eqs. (1.3) and (1.25), the solution to Eq. (1.49) is given by

$$q_r(t) = \frac{\dot{q}_r(o)}{\omega_r} \operatorname{Sin} \omega_r t + q_r(o) \operatorname{Cos} \omega_r t$$

$$+ \frac{1}{m_r^* \omega_r} \int_0^t p_r^*(\tau) \operatorname{Sin} \omega_r(t - \tau) \, d\tau \tag{1.51}$$

where $q_r(o)$ and $\dot{q}_r(o)$ are the initial displacement and initial velocity of the r-th modal coordinate, which are determined from

$$\{q(o)\} = [M^*]^{-1}[\phi]^T[M]\{v(o)\} \tag{1.52a}$$

$$\{\dot{q}(o)\} = [M^*]^{-1}[\phi]^T[M]\{\dot{v}(o)\} \tag{1.52b}$$

in which

$$[M^*] = \begin{bmatrix} m_1^* & & 0 \\ & m_2^* & \\ 0 & & m_n^* \end{bmatrix}$$

The dynamic response of the given system is then obtained by solving for each modal coordinate separately and superposing each solution according to Eq. (1.46). This procedure is also called the *mode superposition method*.

Damped System

The governing equation of motion is given in Eq. (1.26) as

$$[M]\{\ddot{v}\} + [C]\{\dot{v}\} + [K]\{v\} = \{P(t)\} \tag{1.53}$$

The modal analysis technique will lead to uncoupled modal equations if the damping matrix is assumed to be proportional to the mass and stiffness matrices in the form

$$[C] = \alpha[M] + \beta[K] \tag{1.54}$$

where α and β are the proportionality constants.

Since the mass and stiffness matrices satisfy the orthogonality condition, pre- and postmultiplying Eq. (1.52) will lead to

$$m_r^* \ddot{q}_r + c_r^* \dot{q}_r + k_r^* q_r = p_r^*(t), \qquad r = 1, 2, \ldots, n \tag{1.55a}$$

or

$$\ddot{q}_r + 2\zeta_r \omega_r \dot{q}_r + \omega_r^2 q_r = \frac{p_r^*(t)}{m_r^*}, \qquad r = 1, 2, \ldots, n \tag{1.55b}$$

where

$$c_r^* = \alpha m_r^* + \beta k_r^* \tag{1.56a}$$

or

$$\zeta_r = \frac{\alpha}{2\omega_r} + \frac{\beta\omega_r}{2} \tag{1.56b}$$

The above equation can be used to determine the values of α and β provided the damping ratios in any two modes are known. If damping ratios are known for the fundamental mode and for a higher mode, the resulting values of α and β will give a reasonable approximation of damping for the intermediate modes. However, often the damping is known only for the fundamental mode from the test conducted on similar structures. Then the accepted procedure is to set $\alpha = 0$ and $\beta = 2\zeta_1/\omega_1$.

According to Eqs. (1.3) and (1.25), the solution to Eq. (1.55) is given by

$$q_r(t) = e^{-\zeta_r\omega_r t}\left[\frac{\dot{q}_r(0) + q_r(0)\zeta_r\omega_r}{\omega_{Dr}} \text{ Sin } \omega_{Dr}t + q_r(0) \text{ Cos } \omega_{Dr}t\right]$$

$$+ \frac{1}{m_r^*\omega_{Dr}}\int_0^t P_r^*(\tau) e^{-\zeta_r\omega_r(t-\tau)} \text{ Sin } \omega_{Dr}(t - \tau) \, d\tau$$

$$r = 1, 2, \ldots, n \quad (1.57)$$

where $\omega_{Dr} = (1 - \zeta_r^2)^{1/2}\omega_r$, and $q_r(0)$ and $\dot{q}_r(0)$ are determined from Eq. (1.52). The modal coordinates determined from Eq. (1.57) are superposed according to Eq. (1.46) to obtain the response of the damped system.

Example 1.3. If the top floor of the three-storey shear building in Example 1.2 (Fig. 1.11) is subjected to the rectangular pulse shown in Fig. 1.12, determine the expressions for displacement of the top floor, neglecting damping. Assume zero initial conditions.

Solution. The governing equation of motion is given by

$$10^3\begin{bmatrix} 2.0 & & \\ & 1.5 & \\ & & 1.0 \end{bmatrix}\begin{Bmatrix} \ddot{v}_1 \\ \ddot{v}_2 \\ \ddot{v}_3 \end{Bmatrix} + 10^3\begin{bmatrix} 3000 & -1200 & 0 \\ -1200 & 1800 & -600 \\ & -600 & 600 \end{bmatrix}\begin{Bmatrix} v_1 \\ v_2 \\ v_3 \end{Bmatrix}$$

$$= \begin{Bmatrix} 0 \\ 0 \\ P(t) \end{Bmatrix}$$

where

$$P(t) = P_0, \quad \text{for } 0 < t \le t_0$$

$$= 0, \quad \text{for } t > t_0$$

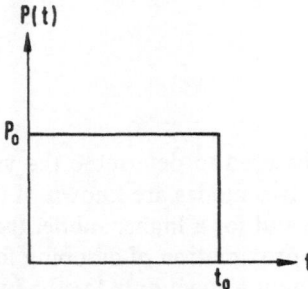

Fig. 1.12. Rectangular pulse.

Expressing the displacement in normal coordinates

$$\begin{Bmatrix} v_1 \\ v_2 \\ v_3 \end{Bmatrix} = \begin{Bmatrix} 0.30 \\ 0.644 \\ 1.0 \end{Bmatrix} q_1(t) + \begin{Bmatrix} -0.676 \\ -0.601 \\ 1.0 \end{Bmatrix} q_2(t) + \begin{Bmatrix} 2.47 \\ -2.57 \\ 1.0 \end{Bmatrix} q_3(t)$$

the governing equation becomes [Eq. (1.49)]

$$\ddot{q}_r(t) + \omega_r^2 q_r(t) = \frac{P(t)}{m_r^*}, \qquad r = 1, 2, 3$$

in which, m_r^*, $r = 1, 2, 3$ are determined from Eq. (1.50a) as

$$m_1^* = 1.80 \times 10^3 \text{ kg}$$

$$m_2^* = 2.46 \times 10^3 \text{ kg}$$

$$m_3^* = 23.11 \times 10^3 \text{ kg}$$

Applying Eq. (1.51)

$$q_r(t) = \frac{P_0}{m_r^* \omega_r^2} (1 - \text{Cos } \omega_r t), \qquad\qquad r = 1, 2, 3, \qquad \text{for } t \le t_0$$

$$= \frac{P_0}{m_r^* \omega_r^2} [\text{Cos } \omega_r(t - t_0) - \text{Cos } \omega_r t], \qquad r = 1, 2, 3, \qquad \text{for } t > t_0$$

Since the mode shapes have unit value at the top floor, the top-floor displacement is obtained as

$$v_3(t) = \sum_{r=1}^{3} q_r(t)$$

$$= P_0 \sum_{r=1}^{3} \frac{1}{m_r^* \omega_r^2} (1 - \text{Cos } \omega_r t), \qquad\qquad t \le t_0$$

$$= P_0 \sum_{r=1}^{3} \left\{ \frac{1}{m_r^* \omega_r^2} [\text{Cos } \omega_r(t - t_0) - \text{Cos } \omega_r t] \right\}, \qquad t > t_0$$

1.4 Continuous System

1.4.1 Equation of Motion

For a uniform beam with distributed mass m per unit length and flexural stiffness EI subjected to an applied loading of $p(z, t)$, the governing equation of motion takes the form

$$m \frac{\partial^2 v}{\partial t^2} + c \frac{\partial v}{\partial t} + EI \frac{\partial^4 v}{\partial z^4} = p(z, t) \tag{1.58}$$

where $v(z, t)$ is the displacement of the beam in the y direction and c the damping coefficient of the viscous dashpots which are distributed uniformly along the beam, connecting the beam to the ground as indicated in Fig. 1.13. Although such an arrangement of dashpots has no physical meaning, it provides a mechanism for energy dissipation.

1.4.2 Free Vibration

The equation of motion for undamped free vibration is given by

$$m \frac{\partial^2 v}{\partial t^2} + EI \frac{\partial^4 v}{\partial z^4} = 0 \tag{1.59}$$

Let

$$v(z, t) = \phi(z) q(t) \tag{1.60}$$

where the function $\phi(z)$ defines the deflected shape as the beam vibrates and $q(t)$ defines the amplitude of vibration.

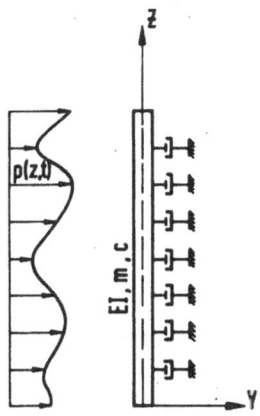

Fig. 1.13. Flexural beam with distributed mass, stiffness and damping.

Substituting Eq. (1.60) into Eq. (1.59) yields

$$m\phi(z)\frac{d^2q(t)}{dt^2} + EIq(t)\frac{d^4\phi(z)}{dz^4} = 0 \tag{1.61}$$

Grouping the terms in z and t together leads to

$$\frac{EI}{m}\frac{1}{\phi(z)}\frac{d^4\phi(z)}{dz^4} = -\frac{1}{q(t)}\frac{d^2q(t)}{dt^2} \tag{1.62}$$

Since the expression on the left is a function of z only while the expression on the right is a function of t only, each expression must be a constant for Eq. (1.62) to be true. Denoting this constant by ω^2, we have

$$EI\frac{d^4\phi(z)}{dz^4} = \omega^2 m\phi(z) \tag{1.63a}$$

$$\frac{d^2q(t)}{dt^2} + \omega^2 q(t) = 0 \tag{1.63b}$$

The solution to Eq. (1.63b) is given by

$$q(t) = A\,\text{Cos}\,\omega t + B\,\text{Sin}\,\omega t \tag{1.64}$$

whereas the general solution to Eq. (1.63a) takes the form

$$\phi(z) = C_1\,\text{Sin}\,\alpha z + C_2\,\text{Cos}\,\alpha z + C_3\,\text{Sinh}\,\alpha z + C_4\,\text{Cosh}\,\alpha z \tag{1.65}$$

in which

$$\alpha^4 = \omega^2 m/EI \tag{1.66}$$

The constants C_1, C_2, C_3 and C_4 depend on the boundary conditions, while the constants A and B depend on the initial conditions. Three of the four constants in Eq. (1.61) and the value of ω are determined from the boundary conditions and hence the modal shape will be known. The constants A and B which absorb the fourth constant of Eq. (1.65) are determined from the initial conditions and hence the amplitude of free vibration can be determined.

Cantilever Beam

For a cantilever beam fixed at $z = 0$ and free at $z = H$, the boundary conditions are

$$v(o, t) = 0, \qquad \frac{dv}{dz}(o, t) = 0 \tag{1.67a}$$

$$EI\frac{d^2v(H, t)}{dz^2} = 0 \qquad EI\frac{d^3v(H, t)}{dz^3} = 0 \tag{1.67b}$$

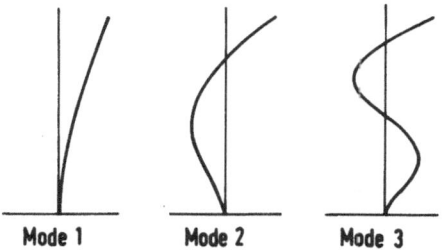

Fig. 1.14. Free vibration mode shapes for a cantilever flexural beam.

representing zero displacement and zero slope at the fixed end, and zero bending moment and zero shear at the free end. Substituting Eq. (1.65) into the boundary conditions leads to the following frequency equation and the mode shape:

$$\text{Cos } \alpha H \text{ Cosh } \alpha H + 1 = 0 \tag{1.68}$$

$$\phi(z) = \left[\text{Sin } \alpha z - \text{Sinh } \alpha z + \frac{\text{Sin } \alpha H + \text{Sinh } \alpha H}{\text{Cos } \alpha H + \text{Cosh } \alpha H} (\text{Cosh } \alpha z - \text{Cos } \alpha z) \right] \tag{1.69}$$

The lowest three values of αH which satisfy Eq. (1.68) are 1.875, 4.694 and 7.855. The first three frequencies are obtained by substituting these values of αH into the following equation:

$$\omega_n = \frac{(\alpha H)^2}{H^2} \sqrt{\frac{EI}{m}} \tag{1.70}$$

The corresponding mode shapes are depicted in Fig. 1.14. They represent the sway modes of towers and tall buildings that deform predominantly in flexural modes.

1.4.3 Orthogonality of Modes of Vibration

If $\phi_r(z)$ and $\phi_s(z)$ are two distinct modes of vibration corresponding to frequencies ω_r and ω_s, then from Eq. (1.59) we have

$$\frac{d^4\phi_r(z)}{dz^4} = \frac{\omega_r^2 m}{EI} \phi_r(z) \tag{1.71a}$$

$$\frac{d^4\phi_s(z)}{dz^4} = \frac{\omega_s^2 m}{EI} \phi_s(z) \tag{1.71b}$$

Multiplying Eq. (1.71a) by $\phi_s(z)$ and Eq. (1.71b) by $\phi_r(z)$, and integrating from 0 to H:

$$\omega_r^2 \frac{m}{EI} \int_0^H \phi_r(z) \phi_s(z) \, \mathrm{d}z = \int_0^H \phi_s(z) \frac{\mathrm{d}^4 \phi_r(z)}{\mathrm{d}z^4} \, \mathrm{d}z \tag{1.72a}$$

$$\omega_s^2 \frac{m}{EI} \int_0^H \phi_r(z) \phi_s(z) \, \mathrm{d}z = \int_0^H \phi_r(z) \frac{\mathrm{d}^4 \phi_s(z)}{\mathrm{d}z^4} \, \mathrm{d}z \tag{1.72b}$$

Subtracting Eq. (1.72b) from Eq. (1.72a) leads to

$$(\omega_r^2 - \omega_s^2) \frac{m}{EI} \int_0^H \phi_r(z) \phi_s(z) \, \mathrm{d}z$$

$$= \int_0^H \left\{ \phi_s(z) \frac{\mathrm{d}^4 \phi_r(z)}{\mathrm{d}z^4} - \phi_r(z) \frac{\mathrm{d}^4 \phi_s(z)}{\mathrm{d}z^4} \right\} \, \mathrm{d}z \tag{1.73}$$

For free, fixed or pinned end conditions, each term on the right-hand sides through integration by parts becomes

$$\int_0^H \phi_s(z) \frac{\mathrm{d}^4 \phi_r(z)}{\mathrm{d}z^4} \, \mathrm{d}z = \int_0^H \frac{\mathrm{d}^2 \phi_s(z)}{\mathrm{d}z^2} \frac{\mathrm{d}^2 \phi_r(z)}{\mathrm{d}z^2} \, \mathrm{d}z \tag{1.74}$$

Thus, for $\omega_r \neq \omega_s$, Eq. (1.73) yields the following orthogonality condition:

$$\int_0^H \phi_r(z) \phi_s(z) \, \mathrm{d}z = 0, \qquad r \neq s \tag{1.75}$$

Substituting Eq. (1.75) into Eq. (1.72a) yields

$$\int_0^H \frac{\mathrm{d}^2 \phi_r(z)}{\mathrm{d}z^2} \frac{\mathrm{d}^2 \phi_s(z)}{\mathrm{d}z^2} \, \mathrm{d}z = 0, \qquad r \neq s \tag{1.76}$$

For a beam with non-uniform mass and stiffness, the corresponding orthogonality conditions are

$$\int_0^H m(z) \phi_r(z) \phi_s(z) \, \mathrm{d}z = 0 \tag{1.77a}$$

$$\int_0^H EI(z) \frac{\mathrm{d}^2 \phi_r(z)}{\mathrm{d}z^2} \frac{\mathrm{d}^2 \phi_s(z)}{\mathrm{d}z^2} \, \mathrm{d}z = 0 \tag{1.77b}$$

1.4.4 Forced Vibration Response

Consider the undamped forced vibration given by

$$m \frac{\partial^2 v}{\partial t^2} + EI \frac{\partial^4 v}{\partial z^4} = p(z, t) \tag{1.78}$$

As the displacement can be expressed as a linear combination of all possible modes, let

$$v(z, t) = \sum_{r=1}^{\infty} \phi_r(z) q_r(t) \tag{1.79}$$

where $\phi_r(z)$ is the r-th mode shape and $q_r(t)$ is the modal coordinate. Substituting Eq. (1.79) into Eq. (1.78) leads to

$$\sum_{r=1}^{\infty} \phi_r(z) \frac{d^2 q_r(t)}{dt^2} + \frac{EI}{m} \sum_{n=1}^{\infty} \frac{d^4 \phi_r(z)}{dz^4} q_r(t) = \frac{p(z, t)}{m} \tag{1.80}$$

Applying Eq. (1.63a), the above expression becomes

$$\sum_{n=1}^{\infty} \phi_r(z) \frac{d^2 q_r(t)}{dt^2} + \sum_{n=1}^{\infty} \phi_r(z) \omega_r^2 q_r(t) = \frac{p(z, t)}{m} \tag{1.81}$$

Premultiplying Eq. (1.81) by $\phi_s(z)$ and integrating from 0 to H:

$$\sum_{r=1}^{\infty} \left(\int_0^H \phi_s(z) \phi_r(z) \, dz \right) \ddot{q}_r(t) + \sum_{r=1}^{\infty} \left(\int_0^H \phi_s(z) \phi_r(z) \omega_r^2 \, dz \right) q_r(t)$$

$$= \frac{1}{m} \int_0^H p(z, t) \phi_s(z) \, dz \tag{1.82}$$

In view of the orthogonality condition given by Eq. (1.75), every term in the summation vanishes, except when $r = s$. Thus the above equation reduces to

$$\ddot{q}_r(t) + \omega_r^2 q_r(t) = \frac{P_r(t)}{M_r}, \qquad r = 1 \text{ to } \infty \tag{1.83}$$

in which the generalized force $P_r(t)$ and the generalized mass M_r are given by

$$P_r(t) = \int_0^H p(z, t) \phi_s(z) \, dz \tag{1.84a}$$

$$M_r(t) = \int_0^H m \phi_s^2(z) \, dz \tag{1.84b}$$

It is to be noted that Eq. (1.83) represents a set of single-degree-of-freedom systems. There are an infinite number of equations for a continuous system. However only the first few equations need to be solved since the contributions from higher modes become negligible. The solution to each single-degree-of-freedom system is obtained as discussed in section 1.2.4. Then the response in physical coordinates is obtained by substituting the modal coordinates in Eq. (1.79).

If the equation of motion contains a damping term as in Eq. (1.58), then if the damping coefficient c is assumed to be proportional to m, the governing equation in modal coordinates becomes

$$\ddot{q}_r + 2\zeta_r\omega_r\dot{q}_r(t) + \omega_r^2 q_r(t) = \frac{P_r(t)}{M_r}, \qquad r = 1 \text{ to } \infty \tag{1.85}$$

where ζ_r is the damping ratio in the r-th mode of vibration. The solution to the above equation is given by

$$q_r(t) = \frac{1}{M_r\omega_{Dr}} \int_0^t P_r(\tau)\, e^{-\zeta_r\omega_r(t-\tau)} \sin \omega_{Dr}(t-\tau)\, d\tau \tag{1.86}$$

in which

$$\omega_{Dr} = \omega_r(1 - \zeta_r^2)^{1/2} \tag{1.87}$$

Substituting Eq. (1.87) into Eq. (1.79) yields the forced vibration response of damped continuous system.

References

1.1. Clough, R W and Penzien J, *Dynamics of Structures*, McGraw-Hill, New York, 1975.
1.2. Meirovitch L, *Elements of Vibration Analysis*, McGraw-Hill, New York, 1975.

Chapter 2

Behaviour of Buildings under Lateral Loads

2.1 Structural Systems

Gravity loads are the primary loading on a building. However, as a building becomes taller, it must have adequate strength and stiffness to resist lateral loads imposed by winds and moderate earthquakes. As the height of a building increases, the additional stiffness required to control the deflection, rather than the strength of the members, dictates the design. Figure 2.1 shows the additional weight of steel required to resist the wind loads as the number of storeys increases. Buildings up to 10 storeys, designed for gravity loads, can resist lateral loads without any increase in the size of members, because of the increase in the permissible stresses allowed for combined loading. Beyond 10 storeys, the additional material required to resist the lateral load increases non-linearly. Thus for reasons of economy it is desirable to use an appropriate structural system to resist the lateral loads, in addition to gravity loads.

A tall building essentially comprises several vertical cantilevers tied together by the floor slabs. Under horizontal loading, each cantilever bends about its own axis, but deforms in unison with other cantilevers owing to the in-plane rigidity of the floor slabs. The various types of vertical cantilever used in

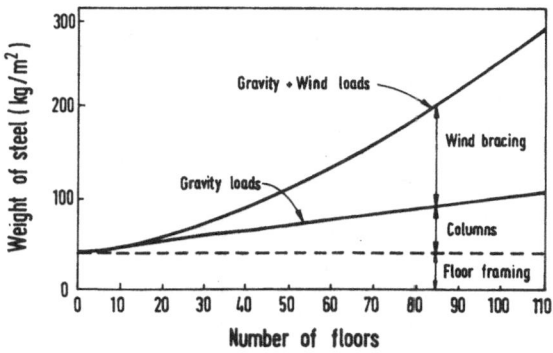

Fig. 2.1. Weight of steel in high-rise buildings.

buildings are rigid frame, braced frame, wall and tube. They individually or in combination form the structural system which resists the lateral loads in a building. The structural systems used in tall buildings are:

1. Braced frame
2. Rigid frame
3. Shear wall
4. Shear wall–frame
5. Framed tube
6. Tube in tube
7. Bundled tube
8. Outrigger-braced.

Many of the above structural systems are suitable for either steel or reinforced concrete buildings, however some are more appropriate for steel and some for reinforced concrete buildings. The behaviour of these structural systems and their areas of application will now be discussed.

2.1.1 Braced Frame Structures

A braced frame consists of columns, beams and diagonal braces, so that the entire system acts as a vertical cantilever truss. The girders and braces form the web of the truss while the columns act as the chord. The horizontal shear due to the lateral load is resisted by the axial action of the braces, and thus it is an efficient system for steel buildings of any height in providing the required stiffness and strength to resist the lateral load.

The common types of braced frame are shown in Fig. 2.2. The axial forces acting in the braces, columns and girders are also shown in this figure. Because the lateral load is reversible, the braces are subjected to both tension and compression, and thus they must be designed for the more stringent case of compression. Consequently, for larger bay widths, K-type bracing is preferred

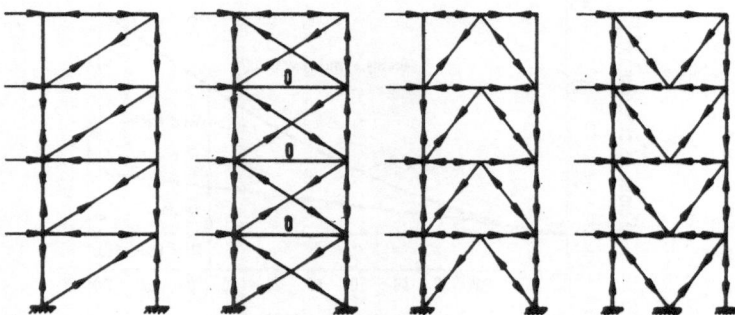

Fig. 2.2. Direction of axial forces in braces, columns and girders of different types of braced frames.

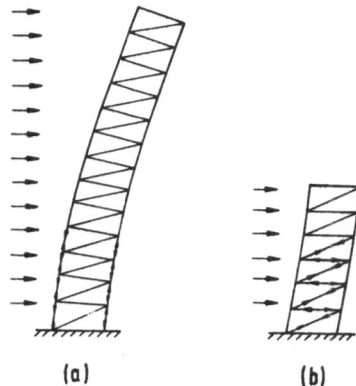

Fig. 2.3. (a) Flexural deflection of high-rise braced frame structure and (b) shear deflection of low-rise braced frame structure.

because of the shorter length of the braces. In resisting the horizontal shear, the diagonals are in tension while the girders are in compression. Thus the elongation of the diagonal and shortening of the girders give rise to shear deformation of the frame. The external moment is resisted by tension in the windward columns and compression in the leeward columns. Consequently, the extension of the windward columns and shortening of the leeward columns tend to cause flexural deformation in the structure. Thus low-rise buildings deflect predominantly in shear mode (Fig. 2.3b) while high-rise buildings deflect predominantly in flexural mode (Fig. 2.3a).

As diagonal braces obstruct the internal planning and locations of door and window openings, they should be located where such access is not required, for example around the elevator, service and stair wells. They may also be placed along walls and partition lines.

2.1.2 Rigid Frame Structures

A rigid frame structure consists of columns and girders connected by moment-resisting joints (Fig. 2.4). Resistance to lateral loading is provided by the bending resistance of the columns, girders and their connections. Rigid frames are economical for buildings up to 25 storeys. For more storeys than this, deeper girders are required to control the drift, and hence the design becomes uneconomical. However, rigid frames can be used economically for taller buildings, in combination with shear walls or braced frames. The open rectangular arrangement of the members of the rigid frame allows freedom for the internal layout. Because of the inherent rigidity of the reinforced concrete joints, the rigid frame construction is ideally suited for reinforced concrete buildings. However, this structural system is also used for steel buildings, although moment-resisting connections in steel are costly.

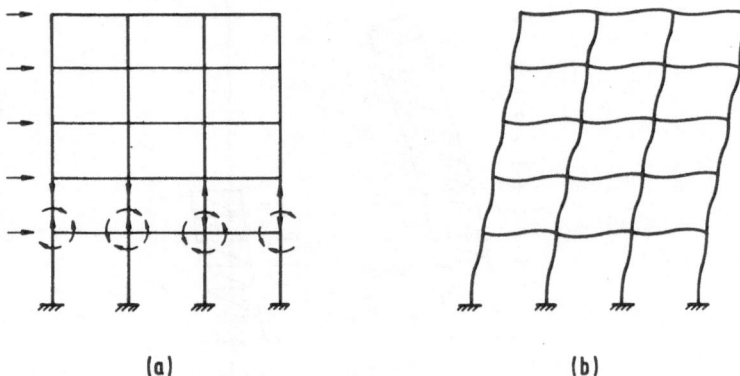

Fig. 2.4. (a) Rigid frame and (b) deflection configuration of rigid frame.

When a rigid frame is subjected to a lateral load, the horizontal shear in a storey is resisted by shear forces in the columns of that storey. The shear causes the columns of the storey to bend in double curvature with points of contraflexure at approximately mid-storey level. The moments acting on joints from columns above and below, as shown in Fig. 2.4a, are resisted by the girders framing into each joint. The corresponding shear causes each girder to bend in double curvature, with points of contraflexure at approximately mid-span. These deformations of columns and girders allow racking of the frame with horizontal deflection in each storey, causing an overall deflection resembling a shear mode, as shown in Fig. 2.4b.

The overall moment caused by the external lateral load is resisted at each storey level by a couple resulting from axial forces in columns, namely tension and compression on opposite sides of the structure. The extension and shortening of the columns cause overall bending, with horizontal deflection of the structure. The storey drift due to overall bending increases with height, while that due to racking tends to decrease. However, the contribution of overall bending to total drift is less than 10% of that of racking, and hence the overall deflection configuration of a rigid frame resembles a shear mode.

2.1.3 Shear Wall Structures

Shear wall structures consist of reinforced concrete vertical walls, in the form of separate planar walls or non-planar assemblies of connected walls around elevator, stair and service shafts (Fig. 2.5). The walls act as cantilever beams which undergo lateral deflection as a result of bending and shear. The ratio of bending deflection to shear deflection of a shear wall increases with the ratio of height to width of the wall. For example, the bending deflection is about 7 times the shear deflection for a height to width ratio of 3. Because of high in-plane stiffness and strength, shear walls are ideally suited to resisting lateral loads. As shear walls are much stiffer horizontally than rigid frames, they are

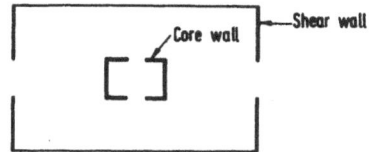

Fig. 2.5. Shear wall structure.

economical up to 35 storeys. However, shear walls of solid form tend to restrict planning where open internal spaces are required.

Shear walls often contain openings to accommodate windows and doors. For example, Fig. 2.6 shows a coupled wall structure where two walls are connected by beams or stiff floor slabs at floor levels. Depending on the stiffness of the connecting members, the two walls can behave independently, each bending about its own axis (for very flexible coupling beams), or as a composite cantilever, bending about a common centroidal axis of the walls. Thus depending on the relative stiffness of the walls and coupling beams, the coupled shear wall structure deflects in either flexural, shear, or a combination of flexural and shear modes. Typical coupled walls deflect in a shear–flexure mode, with the flexural mode configuration in the lower region and the shear mode configuration in the upper region.

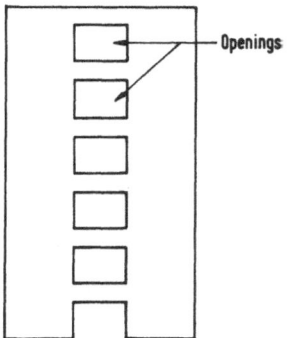

Fig. 2.6. Coupled shear wall structure.

2.1.4 Shear Wall–Frame Structures

An economical way to stiffen a reinforced concrete rigid frame structure is by combining it with a reinforced concrete shear wall. As illustrated in Fig. 2.7, under lateral loads, the rigid frame tends to deflect in a shear mode while the wall deflects in a flexural mode. Because of the horizontal rigidity of the girders and slabs which connect the walls and frames, they are constrained to deform in a common deflected shape, namely the shear–flexure mode, with a flexural mode in the lower region and a shear mode in the upper region. From the

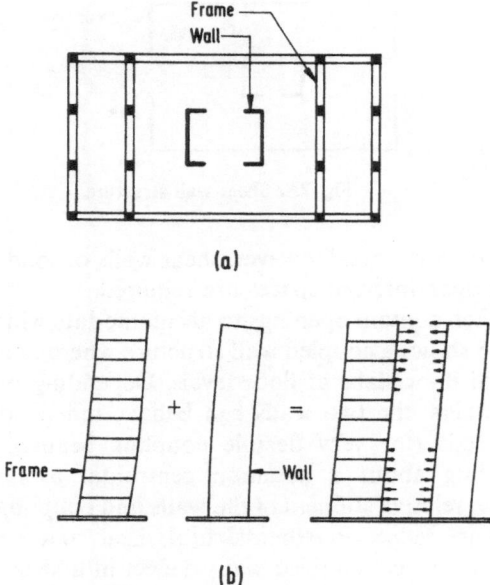

Fig. 2.7. (a) Shear wall–frame structure and (b) deflection configuration of shear wall–frame structure.

directions of the interaction forces, it can be seen that the wall supports the frame in the lower region while the frame supports the wall in the upper region. In consequence, a stiffer and stronger structural system evolves. This structural system can be used for up to 60 storeys. Because of the interaction, the lateral loads cannot be distributed between the walls and frames according to their relative stiffnesses. It is important to note that the total shear carried by the frame at the top storeys can exceed the applied storey shear at these levels, since the frame, in addition to resisting the external lateral load, supports the wall. Thus distributing the applied shear forces according to the relative stiffness can lead to erroneous results. In the case of steel structures, the steel-braced frame takes the role of the shear wall in deflecting in a flexural mode and interacting with the steel rigid frame, which deforms in a shear mode, as illustrated in Fig. 2.8.

2.1.5 Framed Tube Structures

The framed tube is one of the most significant modern developments in the structural forms of tall buildings. In a framed tube structure, lateral resistance is provided by a very stiff moment-resisting frame that forms a tube around the perimeter of the building. The tube is formed by closely spaced columns at 2 to 4 m spacing tied together by deep spandrel beams 1 to 1.5 m deep (Fig. 2.9a) to create a rigid wall-like structure around the building exterior.

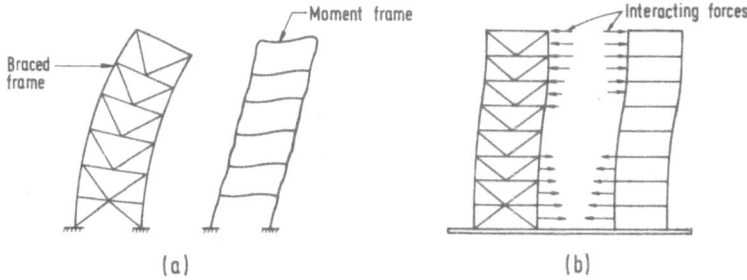

Fig. 2.8. Interaction between braced frame and rigid frame.

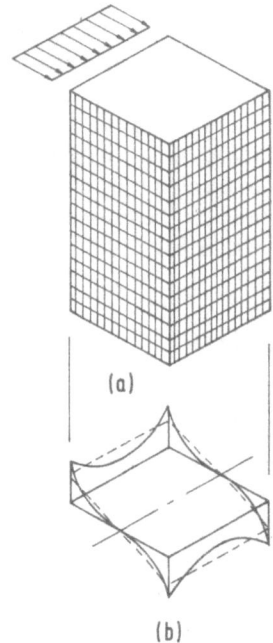

Fig. 2.9. (a) Framed tube structure and (b) axial force distribution in the exterior columns.

Under lateral loads, the framed tube responds as a cantilever box beam; the perimeter frames aligned in the direction of loading act as the webs, and those normal to the direction of the load act as the flanges of the box beam. The entire lateral load is resisted by the exterior tube and the interior columns are assumed to carry only vertical loads. As the internal shear walls or bracings are eliminated, the frame tube structure provides a free space suitable for architectural planning. Furthermore, the closely spaced facade grid can provide mullions for the glazing. The framed tube structure is economical up to 80 storeys for steel buildings and up to 60 storeys for concrete buildings.

Trussed Tube

In practice, the framed tube does not deform in a pure flexural mode, as the shear in the web frame causes bending of the columns and spandrel beams, resulting in racking of the frame which produces shear lag. As a result, the columns in the corners are forced to take a greater share of the load than the columns in between (Fig. 2.9b), and the shear mode deformation becomes more significant in the total deflection configuration. This reduces the efficiency of framed tube structures. As the inherent weakness of the framed tube is due to the flexibility of the spandrel beams, by transferring the shear through diagonal members instead of spandrel beams, the rigidity of the framed tube can be greatly improved. Such a system is called *trussed tube* (Fig. 2.10), which exhibits nearly pure flexural behaviour. The diagonals together with the spandrel beams provide a wall-like rigidity against lateral loads. The diagonals, in addition to carrying a major portion of the lateral loads, act as inclined columns supporting the gravity loads, with nett compressive loads in the diagonals. The dual function of the diagonal members make the trussed tube system very efficient for steel buildings up to 100 storeys. In this arrangement, the column spacing can be much larger than that in framed tube structures. In the case of reinforced concrete buildings, the diagonals can be created by filling the window openings in a diagonal pattern, as shown in Fig. 2.11.

2.1.6 Tube in Tube Structures

An alternative method of stiffening framed tube structures is by introducing interior cores within the building. The lateral loads are now resisted by both the internal core and the exterior framed tube. The rigid floor diaphragm ties the exterior and interior tubes together so that both tubes respond as a single unit to lateral loads. The interaction behaviour between the framed tube and the interior tube is similar to that of a shear wall–frame structure, however the framed tube is much stiffer than the rigid frame, with a shear–flexure mode as

Fig. 2.10. Trussed tube in steel.

Fig. 2.11. Trussed tube in reinforced concrete.

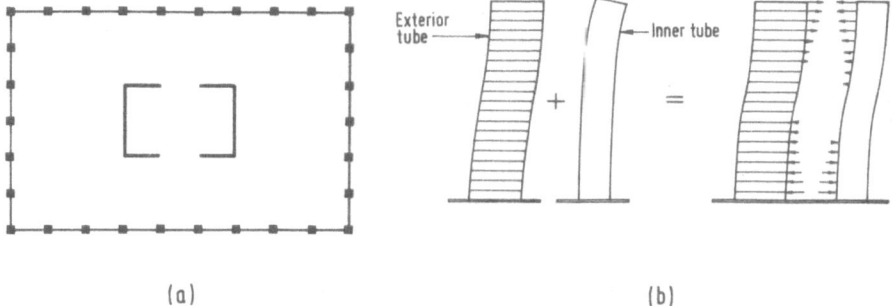

(a)   (b)

Fig. 2.12. (a) Tube in tube structure and (b) deflection configuration of tube in tube structure.

shown in Fig. 2.12. It is seen that the exterior tube resists most of the lateral loads in the upper region of the building, while the interior tube carries most of the lateral loads in the lower region of the building. This system has been used for buildings with large plan areas, with the central cores being used as the service shafts. The column-free spaces between the tubes are ideal for architectural planning.

2.1.7 Bundled Tube Structures

In this arrangement the exterior framed tube is stiffened by interior cross diaphragms made of closely spaced columns tied by spandrel beams (Fig. 2.13), to form an assemblage of several tubes. The interior diaphragms in the direction of the lateral loads act as webs of a huge cantilever beam in resisting the shear forces, thus minimizing the shear lag effects. Otherwise most of the exterior flange columns towards the centre of the building will be of little use in resisting

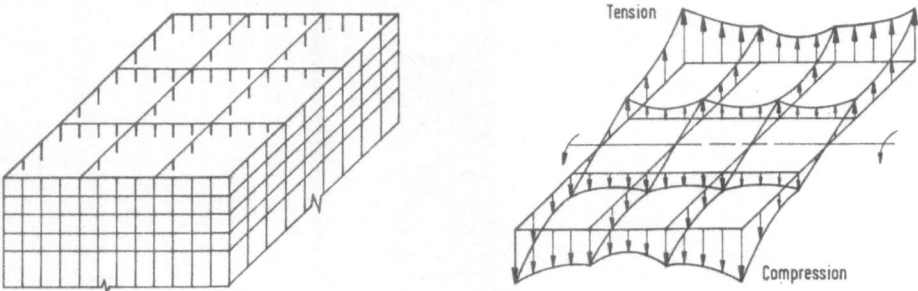

Fig. 2.13. Bundled tube structure.

the overturning moment caused by the lateral loads. The internal diaphragms tend to distribute the axial stresses equally along the flange frames. Although the shear lag effects may still be present, deviation from ideal tube behaviour is significantly less than if there were no internal diaphragms present. The decrease in shear lag effects improves the bending and torsional warping behaviour of the building. The torsional resistance of the bundled tube is much higher than a corresponding single perimeter framed tube system.

The bundled tube system is ideal for buildings which are very tall with extremely large floor areas. It has been used for buildings up to 120 storeys. The individual tubes can be discontinued at any level without loss of structural integrity. This feature enables an architect to create setbacks in a variety of shapes and sizes.

2.1.8 Outrigger-braced Structures

In this structural system, the central core is connected to the perimeter columns by rigid horizontal cantilever 'outrigger' girders or trusses as shown in Fig. 2.14. Under lateral loads, the rotation of the central core in the vertical plane

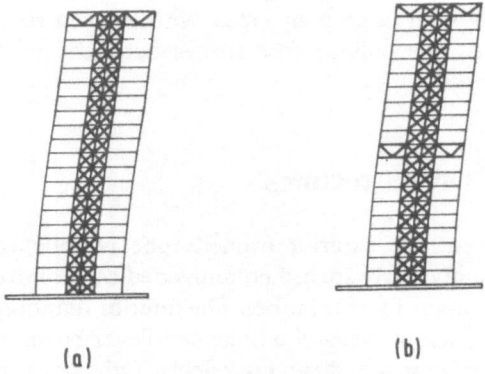

(a)   (b)

Fig. 2.14. Outrigger-braced structure: (a) single outrigger, (b) multiple outriggers.

is restrained through the outrigger by the axial forces in the perimeter columns. As the columns and core behave as a composite structure, the lateral stiffness of the building is increased and hence the lateral deflection and moment in the core are reduced. For this system to be effective, the depth of the outrigger needs to be one to two storeys high. Thus the outriggers are located at the plant levels of the building. By providing a rigid horizontal truss or girder around the perimeter, at the outrigger level, even the columns other than those directly connected to the outrigger can be made to participate in providing the lateral stiffness. Furthermore, a multilevel outrigger structure is more effective than a single-level outrigger structure. This structural system has been used for buildings up to 70 storeys.

2.2 Modelling of Structural Systems

A building consists of structural components (beams, columns, braces, slabs, walls and core walls) and non-structural components (claddings, partitions and stairs). For lateral load analysis, only the principal structural components are included. The effects of other structural components and the non-structural components are assumed to be small and are thus neglected. The principal structural components are identified according to the contribution of each element to the dominant mode of deformation of the building.

2.2.1 Behaviour of Buildings

Horizontal loads due to winds and earthquakes exert a shear, moment and, sometimes, a torque at each level of a building. The building, which behaves as a vertical cantilever, resists the external moment through flexural and axial actions of its vertical components. If the vertical shearing stiffness of the elements connecting the vertical components (girders, slabs, vertical diaphragms and braces) is large, then a greater portion of the external moment will be carried by the axial forces in the vertical components. The horizontal shear at any level of the building is resisted by shear in the vertical components and by the horizontal components of the axial force in any inclined brace members at that level. The torsion is resisted by shear in the vertical components, by the horizontal components of the axial force in any brace members in the vertical plane, and by the torsional resistance of the corewalls. In order to model the torsional resistance of the building correctly, the individual components with appropriate torque constants must be correctly located, and the horizontal shear connection between the components must be correctly modelled.

If a building contains orthogonally connected bents or walls, then the vertical shear connection must be included in the model to obtain the correct bending and torsional resistance of the building. When the vertical components have dissimilar lateral deflection characteristics, there will be a horizontal force interaction between the vertical components. For instance, wall and frame have dissimilar deflection characteristics and when they are connected horizontally by the floor slabs, they are subjected to horizontal interactive forces while assuming a common deflection configuration. Similarly, in constraining different vertical components to have equal twist at each floor level, the floor slabs are subjected to horizontal interaction forces which redistribute the torque between the vertical components.

For lateral load analysis, the floor slabs are assumed to be rigid in their plane. Thus the displacements of all the vertical components at floor level are defined by the rigid body translation and rotation of the floor slabs. The transverse bending stiffness of the slabs in a rigid frame can be neglected, but its value in a flat plate structure must be taken into account. The stiffness of the shear walls about the minor axis and the torsional stiffness of the columns, beams and plane walls are assumed to be negligible. Since beams are monolithic with the floor slabs, and because of the in-plane rigidity of the latter, the axial deformation of the beams is negligible. Furthermore, shear deformation of the beams is negligible unless they are deep. For buildings with a height to width ratio of less than 5, the axial deformation of the columns is negligible.

For an accurate determination of deflection and member forces, a three-dimensional analysis of the structure discretized into finite elements is required. In this model, the beams and columns are discretized into beam elements, braces into truss elements, and the shear walls and core walls into plane stress membrane elements. The degrees of freedom of beam elements, truss elements and membrane elements, for two- and three-dimensional analyses, required for the above discretization are shown in Fig. 2.15. Analysis programs for these elements are available in various commercial packages [2.1, 2.2].

In situations where the transverse bending stiffness of the slabs can be omitted (for example, in beam–column framed structures), the role of the slabs is to serve as rigid diaphragms which distribute the horizontal loadings to the

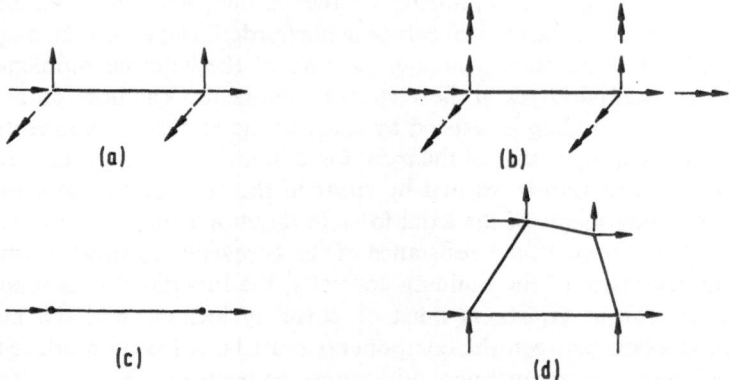

Fig. 2.15. Degrees of freedom of: (a) beam element for two-dimensional analysis, (b) beam element for three-dimensional analysis, (c) truss element, (d) quadrilateral membrane element.

vertical components, and hold the building in plan shape when it translates and twists. Thus in three-dimensional analysis, the in-plane rigidity of the slab may be represented by a horizontal frame of rigid beams joining the vertical components. The beams are rigid for bending in the horizontal plane. However, if the transverse bending stiffness of the slab is part of the lateral load-resisting system, as in flat slab structures, the bending action of the slab between the vertical components is simulated by a connecting beam of equivalent flexural stiffness in the vertical plane. In the horizontal plane, the beam is assumed to be rigid to hold the plan shape of the building.

2.2.2 Modelling of Plane Structures

Discretization of a plane structure is shown in Fig. 2.16. If the span to depth ratio of the beams is greater than 5, shear deformation needs to be included. By setting a large value for the axial area of the beam, the effects of in-plane rigidity of the floor slabs can be simulated. From this analysis, axial forces, shear forces and bending moments in the members can be determined, in addition to vertical and horizontal nodal displacements, and rotation of nodes in the vertical plane.

Generally, rectangular plane stress membrane elements are used for the walls. When a finer mesh is required at a particular region of the wall, quadrilateral elements can be used for the transition region (Fig. 2.17). The membrane plane stress elements yield the horizontal and vertical nodal displacements, vertical and horizontal direct stresses and shear stresses in the elements. As membrane elements do not have a degree of freedom to represent the in-plane rotation of the nodes, when a beam is connected to the wall as in Fig. 2.17, it is effectively connected through a hinge. Thus in order to simulate a rigid connection between the beam and the wall so that a moment can be transferred, a flexurally rigid fictitious beam needs to be added. The beam is rigid for bending in the vertical plane. Alternative ways of placing this fictitious beam are shown in Fig. 2.17. The rotation of the wall is now defined by the relative transverse displacements of the ends of the fictitious beam. The edge of the external beam

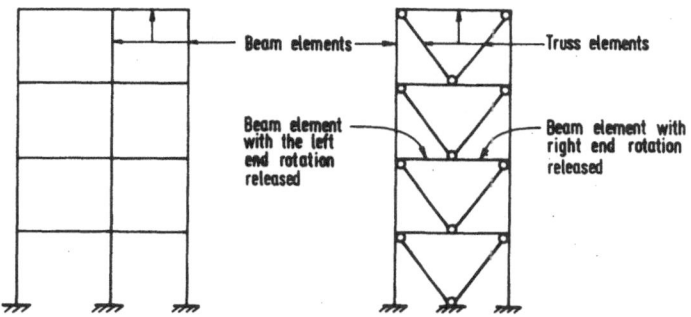

Fig. 2.16. Discretization of rigid frame and braced frame.

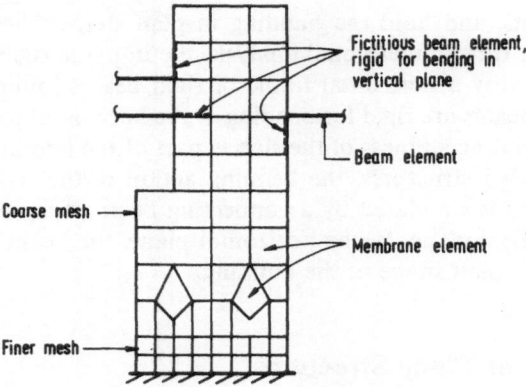

Fig. 2.17. Discretization of plane shear wall.

will be constrained to rotate through this angle and consequently a moment is transferred to the external beam.

2.2.3 Modelling of Three-dimensional Structures

A three-dimensional rigid frame structure, such as that in Fig. 2.18, can be discretized into the beam elements shown in Fig. 2.15b. The element has six degrees of freedom at each node, namely axial, two transverse displacements and rotation about three axes. Thus axial area, shear areas in two transverse directions and moments of inertia about three axes (for in-plane bending, out-of-plane bending and twist) need to be defined. However, in general the axial deformation of the beams and shear deformations of the beams and columns are negligible.

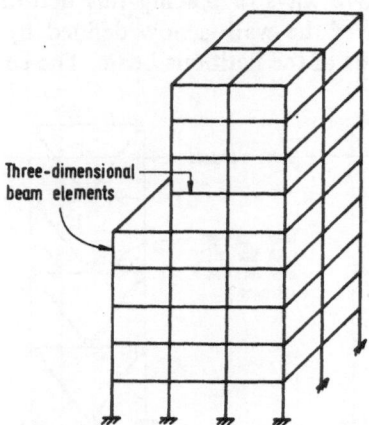

Fig. 2.18. Discretization of three-dimensional rigid frame.

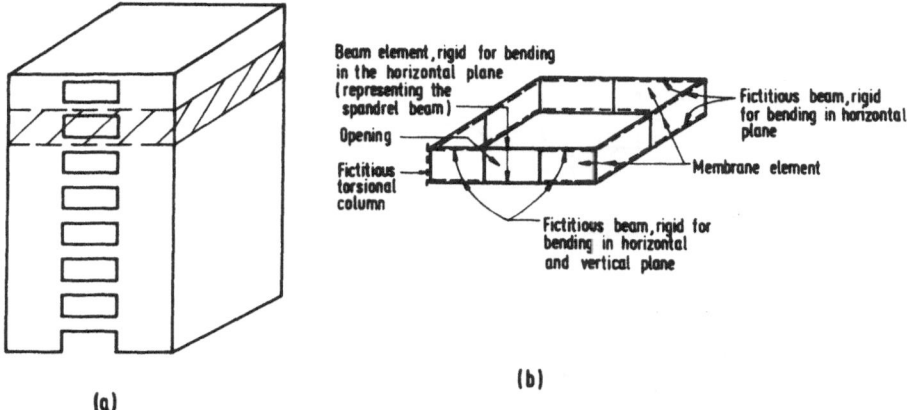

Fig. 2.19. Discretization of three-dimensional shear wall: (a) core wall with opening, (b) discretization of the shaded area.

The three-dimensional shear walls (Fig. 2.19) are assemblies of planar walls and thus each planar wall can be modelled using plane stress membrane elements. However, as the plane stress element cannot provide the out-of-plane rigidity required to maintain the sectional shape of the cross-section, a fictitious horizontal frame of rigid beams needs to be added at each nodal level (Fig. 2.19). The beams are rigid for bending in the horizontal plane. A large axial area and moment of inertia corresponding to bending in the horizontal plane are assigned to the beams in order to simulate the rigid diaphragm effects of the slab.

The torsional stiffness of the wall is simulated by adding a fictitious column as shown in Fig. 2.19. The torsional constant of this column is equal to the sum of the torsional constants of individual plane walls ($\sum \frac{1}{3}bt^3$, where b is the breadth and t the thickness of each plane wall). The axial area and moment of inertia of this fictitious column are taken to be zero. This fictitious column is important for open-section shear wall assemblies, but may be neglected for closed sections.

2.2.4 Reduction of Size of Model

Symmetry

The size of the model can be reduced drastically by taking advantage of geometric symmetry. For instance, a building symmetrical about the Y–Z plane (Fig. 2.20a) when subjected to lateral loads which are antisymmetrical with respect to the Y–Z plane can be analysed by considering only half the structure, provided that the boundary conditions at the plane of symmetry are simulated properly. The appropriate boundary conditions for the structure to deform antisymmetrically about the Y–Z plane are:

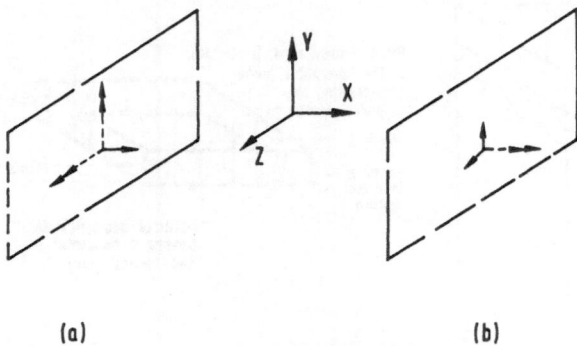

(a)   (b)

Fig. 2.20. Degrees of freedom for: (a) antisymmetric deformation, (b) symmetrical deformation.

1. displacements in the Y–Z plane are zero

2. rotation normal to the Y–Z plane is zero.

If the loads are symmetrical about the plane of symmetry, then the structure will deform symmetrically about the plane of symmetry, and in this case the appropriate boundary conditions are (Fig. 2.20b):

1. displacements normal to the plane of symmetry (the Y–Z plane) is zero

2. rotations in the plane of symmetry are zero.

In the half model, the rigidities of members lying in the plane of symmetry must be halved.

Now consider a symmetrical shear wall–frame subjected to lateral loads (Fig. 2.21). The load is divided into a symmetrical load and an antisymmetrical load in Figs. 2.21b and 2.21c respectively. In Fig. 2.21b when axial deformations are neglected, no forces will be induced in the structure except for the axial load in the beams. Thus, only the structure in Fig. 2.21c needs to be analysed.

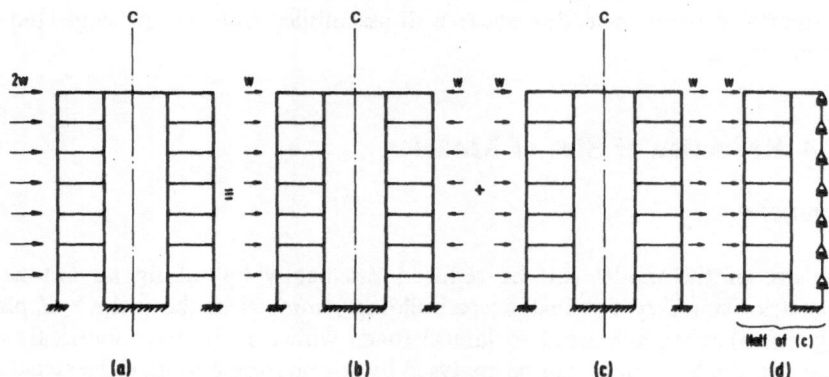

(a)   (b)   (c)   (d)

Fig. 2.21. (a) Symmetrical frame subjected to lateral loads, (b) symmetrical frame subjected to symmetrical loads, (c) symmetrical frame subjected to antisymmetrical loads and (d) equivalent model for frame subjected to antisymmetrical loads.

Taking advantage of antisymmetry, the appropriate model for the analysis is shown in Fig. 2.21d.

Horizontal Lumping

The size of the model may be reduced under certain circumstances. For instance, symmetrical structures under antisymmetrical loading will not twist, and thus a two-dimensional analysis is possible provided the interaction effects between various vertical components are incorporated. Consider the symmetrical building in Fig. 2.22a. It has two identical frames and two identical walls symmetrically situated with respect to the line of symmetry. Both the walls can be lumped into a single wall as they are of the same kind in terms of deformation characteristics. Similarly, the two frames can be lumped into a single frame for the same reason. The properties of the lumped wall and the lumped frame are as indicated in Fig. 2.22b. The lumped frame and the lumped wall are then assembled as a planar model using rigid links (axially stiff truss elements) which play the role of a rigid slab in forcing the lateral displacement of the walls and frames to be the same at each floor level. The model is analysed for total lateral load and the resulting forces in the lumped wall are distributed to the individual walls according to the stiffnesses of the walls. Similarly the resulting forces in the lumped frame are distributed to the individual frames according to the stiffnesses of the frames. Alternatively, taking advantage of antisymmetry, only half the structure need be analysed for half the load using the model in Fig. 2.22c.

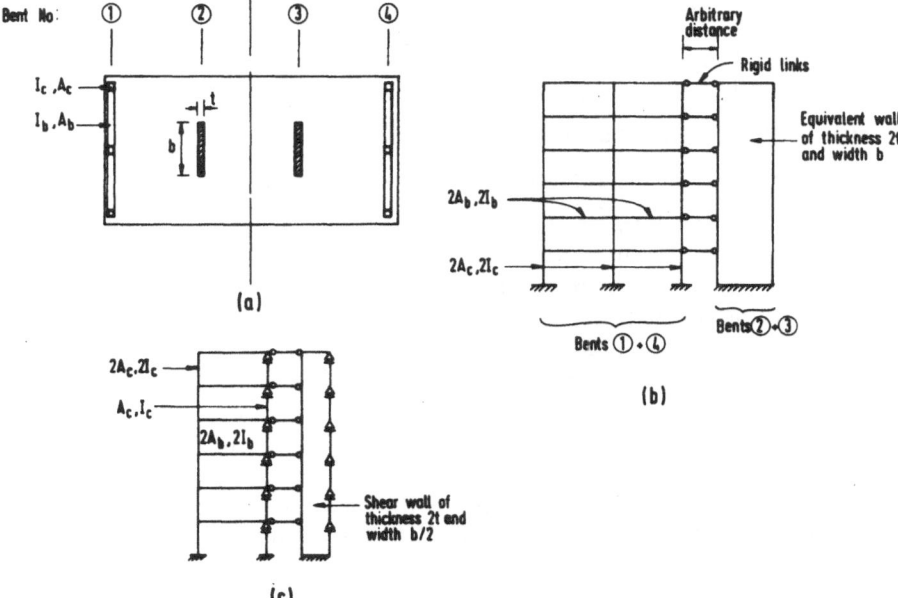

Fig. 2.22. (a) Symmetrical structure with parallel bents, (b) equivalent lumped model and (c) equivalent lumped model for antisymmetric structure.

Vertical Lumping

In tall buildings when storey heights and beam sizes are repetitive, it is possible to lump the floor beams vertically. For instance, for the rigid frame structure in Fig. 2.23a, three floor beams are replaced by a single lumped beam at the middle beam location. The top few and bottom few beams are left in their original positions in order to retain the local effects. The lateral loads are also lumped and applied at the middle beam level (Fig. 2.23b). The properties of the lumped structure are determined as follows:

1. The flexural rigidity of the lumped beam is equal to the sum of the individual beams.
2. The flexural rigidity of the column in the lumped model is obtained by forcing the column to beam stiffness ratio to be the same in both the lumped and unlumped models. If

I_c, I_b = moments of inertia of column and beam in the unlumped model
I'_c, I'_b = moment of inertia of column and beam in the lumped model
l = length of beam in the unlumped and lumped models
h = length of column in the unlumped model
h' = length of column in the lumped model

then equating the stiffness ratio in the lumped model to that in the unlumped model:

$$\frac{(I'_c/h')}{(I'_b/l)} = \frac{(I_c/h)}{(I_b/l)}$$

$$I'_c = \left(\frac{I'_b}{I_b}\right)\left(\frac{h'}{h}\right)I_c$$

If $h' = 3h$, $I'_b = 3I_b$, then $I'_c = 9I_c$

3. The axial areas of the columns in the two models are the same.

In the case of wall–frame structures (Fig. 2.24a), the properties of the wall in the lumped model will be the same as in the unlumped model (Fig. 2.24b), because the wall behaves predominantly in a single curvature.

The resulting moment and shear in a lumped beam must be divided by the number of beams being lumped to obtain the moment and shear in the middle beam of the unlumped model. Once the forces in all the middle beams of the unlumped model are obtained, the forces in the other beams may be obtained by interpolating between the values obtained for the middle beams above and below. The external shear at any level of the unlumped model will be distributed among the columns in the same ratio as the distribution of the shears at the corresponding level of the lumped model. The product of the shear in a particular column and half-storey height (on the assumption that the point of contraflexure is at mid-storey) gives approximately the moment in that column. However, the moment in the wall at any level of the unlumped model is given by the moment in the wall at the corresponding level of the lumped model.

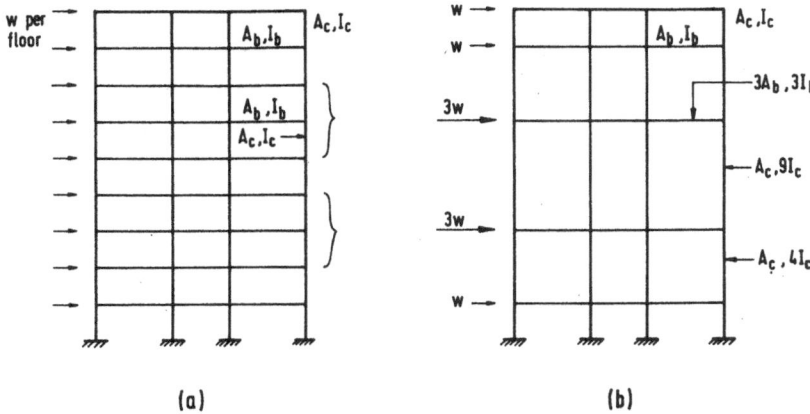

Fig. 2.23. (a) Rigid frame with repetitive floor beams and (b) equivalent lumped model.

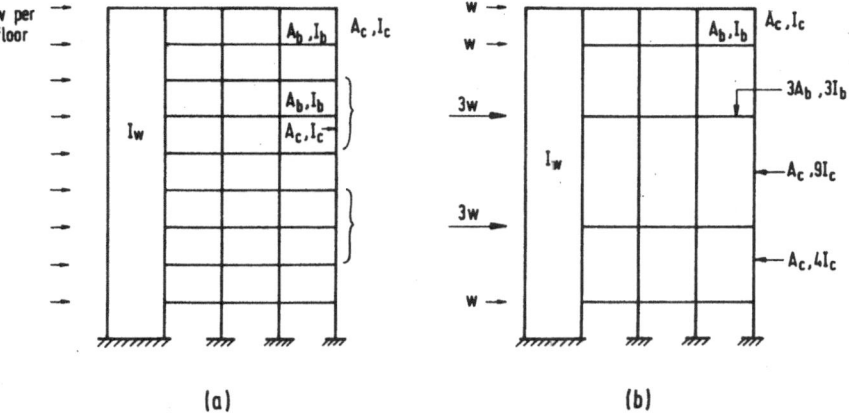

Fig. 2.24. (a) Shear wall–frame structure with repetitive floor beams and (b) equivalent lumped model.

References

2.1. *GTSTRUDL. Georgia Tech structural design language.* Georgia Institute of Technology, Atlanta, Georgia, USA, 1988.

2.2. *ETABS. Three-dimensional analysis of building systems.* Computers and Structures Inc., Berkeley, California, 1989.

Chapter 3

Dynamic Effects of Winds on Buildings

Wind is the movement of air caused by thermal and pressure conditions in the atmosphere. As air moves over the surface of the earth, it strikes and sweeps past all kinds of obstacles in its path, including engineering structures. In many instances, the forces induced and the resulting responses of the structure must be considered if the safety and serviceability of a given design are to be maintained.

Information on the characteristics of winds, which are required to determine the wind loading on objects, is provided by meteorologists, whereas estimation of loading on obstacles due to a defined flow of air is dealt with by aerodynamicists. Structural dynamicists estimate the response of structures from information provided by meteorologists and aerodynamicists. In some instances, additional loading will be induced by deformation of the structures. In such cases, the extra loading arising from wind-induced structural vibrations must be taken into account in the design through the application of the laws of aeroelasticity. The steps involved in the design of structures for wind effects, as illustrated schematically in Fig. 3.1, will be discussed in this chapter with respect to buildings and towers. The wind effects on other engineering structures, such as masts, bridges, chimneys, etc., are beyond the scope of this book.

3.1 Characteristics of Wind

3.1.1 Mean Wind Speed

The velocity of wind (wind speed) at great heights above the ground is constant and is called the *gradient wind speed* $\bar{U}_g$. As shown in Fig. 3.2, close to the ground surface, the wind speed is affected by frictional forces caused by the terrain, and thus there is a boundary layer within which the wind speed varies from zero to the gradient wind speed. The thickness of the boundary layer H_g (gradient height) depends on the ground roughness. For example, the value of

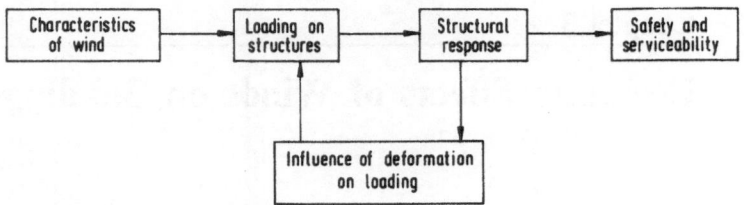

Fig. 3.1. Schematic diagram for design of structures for wind effects.

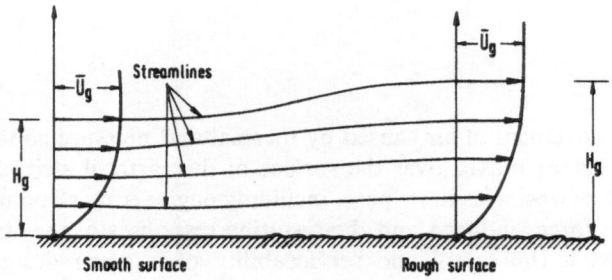

Fig. 3.2. Change of wind speed distribution with terrain roughness.

H_g is 457 m for large cities, 366 m for suburbs, 274 m for open terrain and 213 m for open sea [3.1].

The velocity of wind averaged over one hour is called the *hourly mean wind speed* $\bar{U}$. The mean wind velocity profile within the atmospheric boundary layer is described by a power law

$$\bar{U}(z) = \bar{U}(z_{\text{ref}})\left(\frac{z}{z_{\text{ref}}}\right)^{\alpha} \tag{3.1}$$

where $\bar{U}(z)$ is the mean wind speed at height z above the ground, z_{ref} the reference height, normally taken to be 10 m, and α the power law exponent.

An alternative description of mean wind velocity is given by the logarithmic law

$$\bar{U}(z) = \frac{1}{k}u_* \ln\left(\frac{z-d}{z_0}\right) \tag{3.2}$$

where u_* is the friction velocity, k von Kármán's [3.2] constant (equal to 0.4), z_0 is the roughness length and d the height of the zero plane (where the velocity is zero) above the ground. Generally, the zero plane is about one or two metres below the average height of buildings and trees providing the roughness. Typical values of z_0, α and d are given in Table 3.1 [3.1, 3.3].

The influence of ground roughness on the mean wind profile is depicted in Fig. 3.3. The roughness affects both the thickness of the boundary layer and the power law exponent. As seen from Fig. 3.3, the thickness of the boundary layer and the power law exponent increase with the roughness of the surface.

Table 3.1. Typical values of terrain parameters z_0, α and d.

	z_0 (m)	α	d (m)
City centres	0.7	0.33	15–25
Suburban terrain	0.3	0.22	5–10
Open terrain	0.03	0.14	0
Open sea	0.003	0.10	0

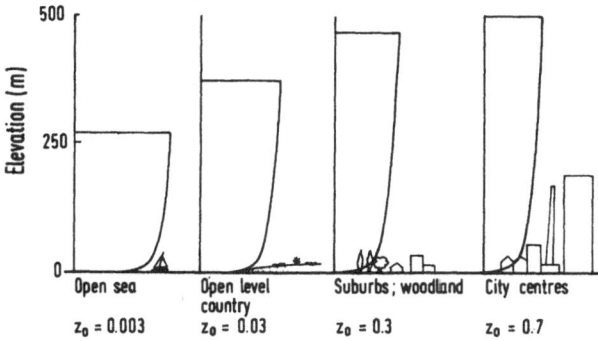

Fig. 3.3. Profiles of mean wind speed within the boundary layer of different terrains.

Consequently the velocity at any height decreases as the surface roughness increases. However, the gradient velocity will be the same for all surfaces. Thus if the velocity of wind for a particular terrain is known, using Eq. (3.1) and Table 3.1 the velocity for some other terrain can be computed.

3.1.2 Turbulence

The variation of wind velocity with time is shown in Fig. 3.4. The eddies generated by the action of wind blowing over obstacles cause the turbulence. In general, the velocity of the wind may be represented in a vector form as

$$U(z, t) = \bar{U}(z)\mathbf{i} + u(z, t)\mathbf{i} + v(z, t)\mathbf{j} + w(z, t)\mathbf{k} \tag{3.3}$$

where u, v and w are the fluctuating components of the gust in the x, y and z axes (longitudinal, lateral and vertical axes) as shown in Fig. 3.5, and $\bar{U}(z)$ is the mean wind along the x axis. The fluctuating component along the mean wind direction, u, is the largest and therefore the most important for vertical structures such as tall buildings which are flexible in the along-wind direction. The vertical component w is important for horizontal structures which are flexible vertically, such as long-span bridges.

An overall measure of the intensity of turbulence is given by the root mean square value (r.m.s.). Thus for the longitudinal component of the

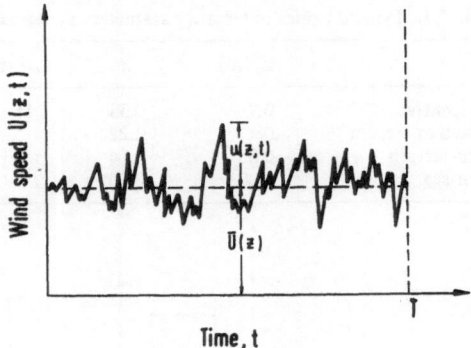

Fig. 3.4. Typical trace of longitudinal component of wind speed with time.

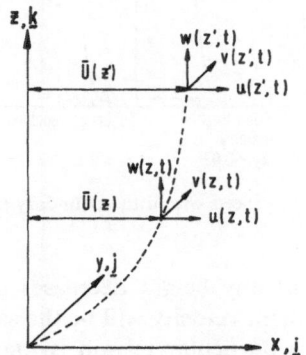

Fig. 3.5. Velocity components of turbulent wind.

turbulence, the r.m.s. value is given by

$$\sigma_u(z) = \left[\frac{1}{T_0} \int_0^{T_0} \{u(z, t)\}^2 \, \mathrm{d}t \right]^{1/2} \tag{3.4}$$

where T_0 is the averaging period. For the statistical properties of the wind to be independent of the part of the record being used, T_0 is taken to be one hour. Thus, over one hour fluctuating wind is a stationary random function.

The value of $\sigma_u(z)$ divided by the mean velocity $\bar{U}(z)$ is called the *turbulence intensity*

$$I_u(z) = \sigma_u(z)/\bar{U}(z) \tag{3.5}$$

which increases with ground roughness and decreases with height. The vertical and lateral turbulence intensities may be similarly defined.

The variance of longitudinal turbulence can be determined from [3.4]

$$\sigma_u^2 = \beta u_*^2 \tag{3.6}$$

Table 3.2. Values of β for various roughness lengths.

z_0 (m)	0.005	0.07	0.30	1.0	2.5
β	6.5	6.0	5.25	4.85	4.0

where u_* is the friction velocity determined from Eq. (3.2) and β, which is independent of height, is given in Table 3.2 for various roughness lengths.

For example if $z = 10$ m, $z_0 = 0.3$ m and $\bar{U}(10) = 18$ m/s, then from Eq. (3.2), assuming the zero plane is 5 m above the ground

$$u_* = \frac{0.4 \times 18}{\ln\left(\dfrac{10-5}{0.3}\right)} = 2.56 \text{ m/s}$$

From Table 3.2, $\beta = 5.25$. Thus, the turbulence intensity

$$I_u(10) = \frac{\sqrt{5.25} \times 2.56}{18} = 0.325$$

3.1.3 Integral Scales of Turbulence

As mentioned previously, the fluctuation of wind velocity at a point is due to eddies transported by the mean wind $\bar{U}$. Each eddy may be considered to be causing a periodic fluctuation at that point with a frequency n. The wavelength of the eddy $\lambda = \bar{U}/n$ is a measure of eddy size. The average sizes of the turbulent eddies are measured by integral length scales. For eddies associated with longitudinal velocity fluctuation u, the integral length scales are L_u^x, L_u^y and L_u^z, describing the size of the eddies in the longitudinal, lateral and vertical directions, respectively. Similarly, there are three integral length scales associated with the lateral and vertical turbulent velocity components v and w.

The dynamic loading on a structure depends on the size of eddies. If L_u^y and L_u^z are comparable to the dimension of the structure normal to the wind, then the eddies will envelope the structure and give rise to well correlated pressures, and thus the effect is significant. On the other hand, if L_u^y and L_u^z are small, then the eddies produce uncorrelated pressures at various parts of the structure and the overall effect of the longitudinal turbulence will be small.

Mathematically the integral length at any height z is obtained from the autocorrelation function of the turbulent velocity. For the longitudinal component of turbulence $u(t)$, the autocorrelation function is determined from

$$R_u(\tau) = \frac{\int_0^{T_0} u(t)u(t+\tau)\,d\tau t}{\int_0^{T_0} u^2(t)\,dt} \tag{3.7}$$

where τ is the time shift of the velocity signal $u(t)$. The plot of autocorrelation function is shown in Fig. 3.6. The average period of eddies is given by the area

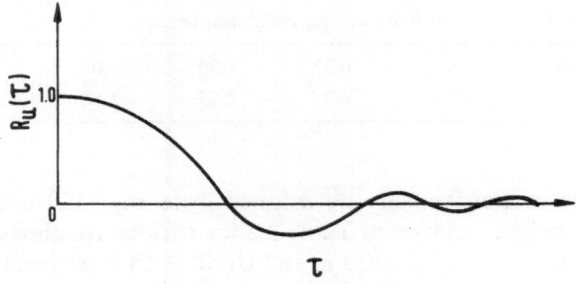

Fig. 3.6. Typical variation of autocorrelation function of wind turbulence.

under the autocorrelation function. Thus

$$T_u = \int_0^\infty R_u(\tau)\, d\tau \tag{3.8}$$

which is the integral time scale. The integral length scale L_u^x is given by

$$L_u^x = \bar{U}(z)/T_u \tag{3.9}$$

where $\bar{U}$ is the mean velocity at the point considered.

The following empirical expression for the integral length scales L_u^x is given by Counihan [3.5]:

$$L_u^x = cz^m \tag{3.10}$$

where L_u^x and z are in metres and the coefficients c and m are given in Fig. 3.7. The expressions for L_u^y and L_u^z are given in [3.6] and [3.7], respectively, as

$$L_u^y = 0.2 L_u^x \tag{3.11}$$

$$L_u^z = 6\sqrt{z} \tag{3.12}$$

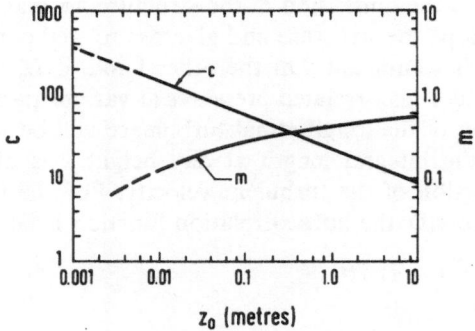

Fig. 3.7. Values of C and m as functions of z_0. (Reproduced by permission of the *Journal of Atmospheric Environment*, p. 888, 1975.).

3.1.4 Spectrum of Turbulence

The frequency content of the turbulence is represented by the power spectrum, which indicates the power or kinetic energy per unit time, associated with eddies of different frequencies. An expression for the power spectrum is given by Simiu [3.8]:

$$\frac{nS_u(z, n)}{u_*^2} = \frac{200f}{(1 + 50f)^{5/3}} \tag{3.13}$$

where $f = nz/\bar{U}(z)$ is the reduced frequency. A typical spectrum of wind turbulence is shown in Fig. 3.8.

The spectrum has a peak value at a very low frequency around 0.04 Hz. As the typical range for the fundamental frequency of tall buildings is 0.1 to 1 Hz, such buildings are affected by high-frequency small eddies characterizing the descending part of the power spectrum.

3.1.5 Cross Spectrum of Turbulence

The cross spectrum of two continuous records is a measure of the degree to which the two records are correlated. If the records are taken at two points M_1 and M_2 separated by a distance r, then the cross spectrum of longitudinal turbulent component is defined as

$$S_{u_1u_2}(r, n) = S^c_{u_1u_2}(r, n) + iS^q_{u_1u_2}(r, n) \tag{3.14}$$

where the real and imaginary parts of the cross spectrum are known as the co-spectrum and the quadrature spectrum, respectively. However, the latter is small enough to be neglected. Thus the co-spectrum may be expressed non-dimensionally as the coherence and is given by

$$\gamma^2(r, n) = \frac{|S_{u_1u_2}(r, n)|^2}{S_{u_1}(n)S_{u_2}(n)} \tag{3.15}$$

where $S_{u_1}(n)$ and $S_{u_2}(n)$ are the longitudinal velocity spectra at M_1 and M_2, respectively. Devenport [3.9] has suggested the following expression for the square root of the coherence:

$$\gamma(r, n) = e^{-f} \tag{3.16}$$

where

$$f = \frac{n[c_z^2(z_1 - z_2)^2 + c_y^2(y_1 - y_2)^2]^{1/2}}{\frac{1}{2}[\bar{U}(z_1) + \bar{U}(z_2)]} \tag{3.17}$$

in which y_1, z_1 and y_2, z_2 are the coordinates of points M_1 and M_2. The line joining M_1 and M_2 is assumed to be perpendicular to the direction of the

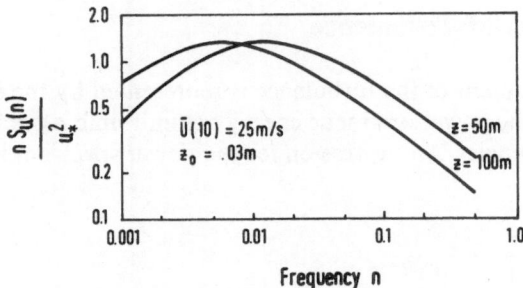

Fig. 3.8. Spectrum of longitudinal turbulence.

mean wind. The suggested values of c_y and c_z for engineering calculations are 16 and 10, respectively [3.10].

3.2 Wind-induced Dynamic Forces

3.2.1 Forces due to Uniform Flow

When a bluff body is immersed in a two-dimensional flow as shown in Fig. 3.9, it is subjected to a nett force in the direction of flow (drag force) and a force perpendicular to the flow (lift force). Furthermore, when the resultant force is eccentric to the elastic centre, the body will be subjected to a torsional moment. For uniform flow, these forces and moment per unit height of the object are determined from

$$F_D = \tfrac{1}{2}\rho C_D B \bar{U}^2 \tag{3.18}$$

$$F_L = \tfrac{1}{2}\rho C_L B \bar{U}^2 \tag{3.19}$$

$$T = \tfrac{1}{2}\rho C_T B^2 \bar{U}^2 \tag{3.20}$$

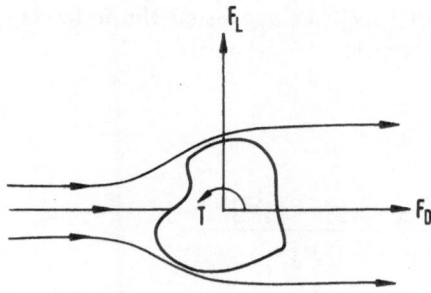

Fig. 3.9. Drag and lift forces and torsional moment on an arbitrary bluff body.

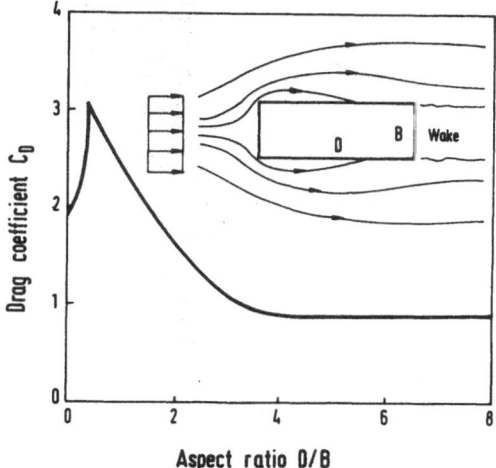

Fig. 3.10. Drag coefficient for a rectangular section with different aspect ratios. (Reproduced by permission of ASCE [3.11]).

where $\bar{U}$ is the mean velocity of the wind, ρ the density of air, C_D and C_L the drag and lift coefficients, C_T the moment coefficient and B the characteristic length of the object, such as the projected length normal to the flow.

The drag coefficient for a rectangular shape is shown in Fig. 3.10 for various depth to breadth ratios [3.11]. The flow separation occurs at windward corners. The shear layers originating from the separation points surround a region known as the *wake*. Near the separation zones, strong shear stresses impart rotational motions to the fluid particles. Thus discrete vortices are produced in the separation layers. For elongated sections, the stream lines which separate at the windward corners re-attach themselves to the body to form a narrower wake. This contributes to the reduction in the drag for larger aspect ratios. For cylindrical shapes, the drag coefficient is dependent on Reynolds number, as indicated in Fig. 3.11.

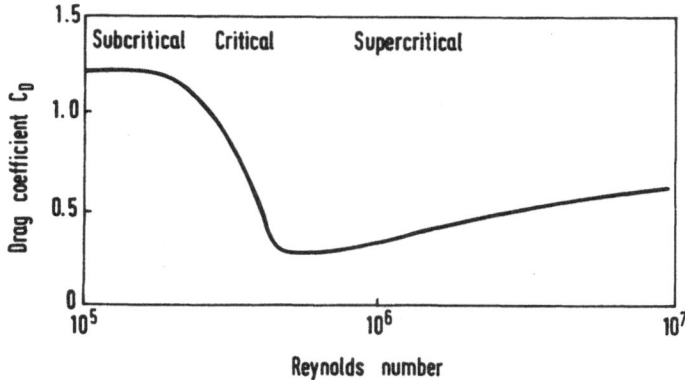

Fig. 3.11. Variation of drag coefficient with Reynolds number for a circular cylinder in uniform flow.

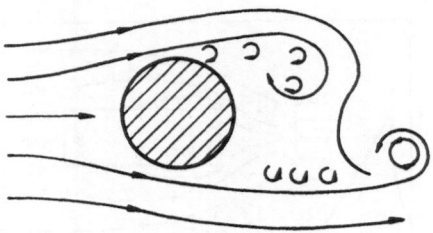

Fig. 3.12. Vortex formation in the wake of a bluff body.

Unlike the drag force, the lift force and torsional moment do not have a mean value for a symmetric object with a symmetric flow around it, as the symmetrical distribution of mean forces acting in the across-wind direction cannot produce a nett force. If the direction of the wind is not parallel to the axes of symmetry or if the object is asymmetrical, then there will be a mean lift force and a torsional moment. However, because of vortex shedding, a fluctuating lift force and a torsional moment will be present in both the symmetric and the non-symmetric structures. Figure 3.12 shows the mechanism of vortex shedding. The air travels over the face of the body until it reaches the points of separation on each side of the body where thin sheets of tiny vortices are generated. As the vortex sheets detach, they interact with one another or roll up into discrete vortices which are shed from the sides of the object. The asymmetric pressure distribution created by the vortices around the cross-section leads to an alternating transverse force (lift force) on the object. The vortex shedding frequency in Hz, n_s, is related to a non-dimensional parameter called the Strouhal number S defined as

$$S = n_s B/\bar{U} \tag{3.21}$$

where $\bar{U}$ is the wind speed and B the width of the object normal to the wind. As shown in the Fig. 3.13, for objects with rounded profiles such as circular

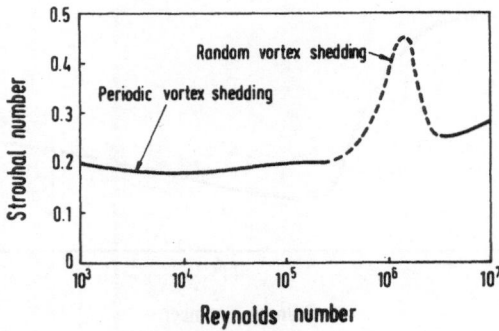

Fig. 3.13. Variation of Strouhal number with Reynolds number for a circular cylinder.

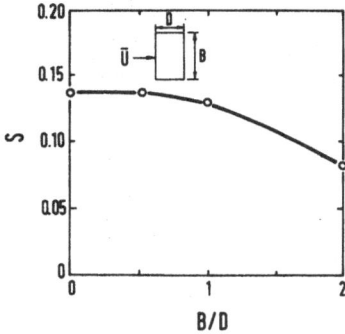

Fig. 3.14. Strouhal number for a rectangular section.

cylinders, the Strouhal number varies with Reynolds number Re defined as

$$Re = \rho \bar{U} B / \mu \qquad (3.22)$$

where ρ is the density of air and μ the dynamic viscosity of the air. The vortex shedding becomes random in the transition region of $4 \times 10^5 < Re < 3 \times 10^6$ where the boundary layer at the surface of the cylinder changes from laminar to turbulent. Outside this transition range, vortex shedding is regular, producing a periodic lift force. For cross-sections with sharp corners, the Strouhal number is independent of the Reynolds number. The variation of Strouhal number with length to breadth ratio of a rectangular cross-section is shown in Fig. 3.14.

3.2.2 Forces due to Turbulent Flow

If the wind is turbulent, then the velocity of the wind in the along-wind direction is described as follows:

$$U(t) = \bar{U} + u(t) \qquad (3.23)$$

where $\bar{U}$ is the mean wind and $u(t)$ the turbulent component in the along-wind direction. The time-dependent drag force per unit height is obtained from Eq. (3.18) by replacing $\bar{U}$ by $U(t)$. As the ratio $u(t)/\bar{U}$ rarely exceeds 0.2 for practical ranges of turbulent intensities, the time-dependent drag force can be expressed as

$$f_D(t) = \bar{f}_D + f_D'(t) \qquad (3.24)$$

where $\bar{f}_D$ and f_D' are the mean and the fluctuating parts of the drag force per unit height which are given by

$$\bar{f}_D = \tfrac{1}{2} \rho \bar{U}^2 C_D B \qquad (3.25)$$

$$f_D' = \rho \bar{U} u C_D B \qquad (3.26)$$

The value of C_D varies slightly because of the presence of turbulence. However, for most cases of interest in practice, drag coefficients obtained under uniform flow can be used in Eqs. (3.25) and (3.26).

The spectral density of the fluctuating part of the drag force is obtained from the autocorrelation function

$$R_{f_D}(\tau) = \lim_{T \to \infty} \frac{1}{T} \int_{-T/2}^{T/2} f'_D(t) f'_D(t + \tau) \, d\tau$$

$$= \rho^2 \bar{U}^2 B^2 C_D^2 \overline{u(t) u(t + \tau)} \tag{3.27}$$

The Fourier transformation of R_{f_D} yields the spectral density as

$$S_{f_D}(n) = 2 \int_{-\infty}^{\infty} R_{f_D}(\tau) \cos 2\pi n\tau \, d\tau$$

$$= \rho^2 \bar{U}^2 B^2 C_D^2 S_u(n) \tag{3.28}$$

where $S_u(n)$ is the spectral density of the turbulent velocity.

From Eq. (3.26), it is evident that the fluctuating drag force varies linearly with the turbulence in the wind buffeting on a structure. Thus, large integral length scale and high turbulent intensities will cause strong buffeting and consequently increase the along-wind response of the structure.

As mentioned previously, the time-varying lift force under uniform flow is due to vortex shedding. The regularity of this vortex shedding is affected by the presence of turbulence in the along wind. Consequently, the across-wind motion and torsional motion due to vortex shedding decrease as the level of turbulence increases.

3.3 Along-wind Response

3.3.1 Point Structures

Elevated water tanks, observation towers, etc. can be classified as point structures where most of the mass is concentrated at a single point. The point structure shown in Fig. 3.15 can be modelled as a single-degree-of-freedom system. The elastic stiffness is provided by the columns supporting the mass. The equation of motion of the system is

$$\ddot{x} + 2\zeta_1(2\pi n_1)\dot{x} + (2\pi n_1)^2 x = F(t)/m \tag{3.29}$$

where x is the displacement, n_1 the natural frequency in Hz, ζ_1 the damping ratio, m the mass and $F(t)$ the fluctuating drag force which according to Eq. (3.26) may be expressed as

$$F(t) = \rho \bar{U} u(t) C_D BD \tag{3.30}$$

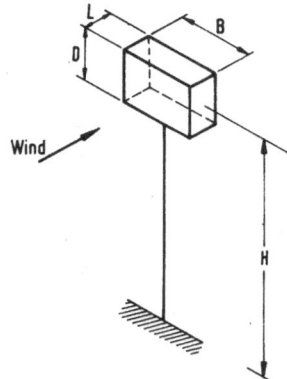

Fig. 3.15. Schematic representation of a point structure.

where B and D are the breadth and depth of the structure as shown in Fig. 3.15. The power spectral density of the fluctuating force, according to Eq. (3.28), takes the form

$$S_F(n) = \rho^2 \bar{U}^2 B^2 D^2 C_D^2 S_u(n) \tag{3.31}$$

where S_u is the power spectral density of the along-wind turbulence. However, in practice, the presence of the structure distorts the turbulent flow, particularly the small high-frequency eddies. A correction factor known as the *aerodynamic admittance function* $\chi(n)$ may be introduced [3.12] to account for these effects. The following empirical formula has been suggested by Vickery [3.13] for $\chi(n)$:

$$\chi(n) = \frac{1}{1 + \left[\dfrac{2n\sqrt{A}}{\bar{U}(z)} \right]^{4/3}} \tag{3.32}$$

where A is the frontal area of the structure. Now with the introduction of the aerodynamic admittance function, Eq. (3.31) may be rewritten as

$$S_F(n) = \rho^2 \bar{U}^2 B^2 D^2 C_D^2 \chi^2(n) S_u(n) \tag{3.33}$$

As the forcing function is random, the response of the system will be non-deterministic. Using random vibration theory [3.14], the power spectrum of the response can be determined as

$$S_x(n) = |H_1(n)|^2 S_F(n) \frac{1}{K^2} \tag{3.34}$$

where $K = (4\pi^2 n_1^2 m)$ is the stiffness and $|H_1(n)|$ is the mechanical admittance function obtained from Eq. (1.11) as

$$|H_1(n)| = \frac{1}{\left(\left[1 - \left(\dfrac{n}{n_1} \right)^2 \right]^2 + 4\zeta_1^2 \left(\dfrac{n}{n_1} \right)^2 \right)^{1/2}} \tag{3.35}$$

The variance of the displacement is obtained from

$$\sigma_x^2 = \int_0^\infty S_x(n) \, dn$$

$$= \frac{1}{K^2} \int_0^\infty |H_1(n)|^2 S_F(n) \, dn \tag{3.36}$$

The calculation of the above integral is very much simplified by observing the plot of the two components of the integrand shown in Fig. 3.16. The mechanical admittance function is either 1.0 or zero for most of the frequency range. However, over a relatively small range of frequencies around the natural frequency of the system it attains very high values if the damping is small. As a result, the integrand takes the shape shown in Fig. 3.16c. It has a sharp spike around the natural frequency of the system. The broad hump is governed by the shape of the turbulent velocity spectrum which is modified slightly by the aerodynamic admittance function. The area under the broad hump is the broad-band or non-resonant response, whereas the area in the vicinity of the natural frequency gives the narrow-band or resonant response. Thus Eq. (3.36) can be rewritten as

$$\sigma_x^2 = \frac{1}{K^2} \int_0^{n_1 - \Delta n} S_F(n) \, dn + \frac{1}{K^2} S_F(n_1) \int_{n_1 - \Delta n}^{n_1 + \Delta n} |H_1(n)|^2 \, dn$$

$$= \frac{1}{K^2} \int_0^{n_1 - \Delta n} S_F(n) \, dn + \frac{\pi n_1 S_F(n_1)}{4 \zeta_1 K^2}$$

$$= \sigma_B^2 + \sigma_D^2 \tag{3.37}$$

where σ_B and σ_D are the variance of the non-resonant and the resonant displacements, respectively.

The r.m.s. acceleration is obtained from

$$\ddot{\sigma}_D = (2\pi n_1)^2 \sigma_D \tag{3.38}$$

3.3.2 Line-like Structures

Tall slender buildings, such as that shown in Fig. 3.17, can be idealized as a line-like structure where the breadth of the structure is small compared with the height. Modelling the building as a continuous system, the governing equation of motion for along-wind displacement $x(z, t)$ can be written as [3.15]

$$m(z)\ddot{x}(z, t) + c(z)\dot{x}(z, t) + EI(z)x''''(z, t) - GA(z)x''(z, t) = f(z, t) \tag{3.39}$$

where m, c, EI and GA are respectively the mass, damping coefficient, flexural rigidity and shear rigidity per unit height. Furthermore, $f(z, t)$ is the fluctuating wind load per unit height given in Eq. (3.26). The dots denote the derivative with respect to time t and the primes denote the derivative with respect to z.

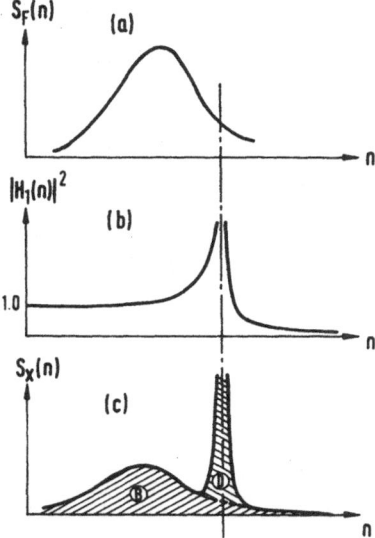

Fig. 3.16. Schematic diagrams for computation of Eq. (3.36).

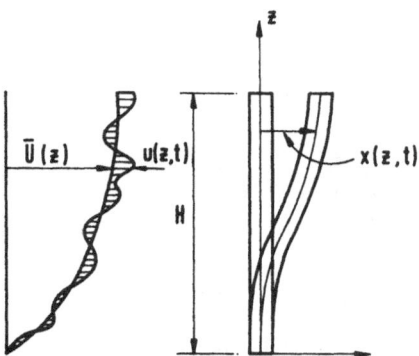

Fig. 3.17. Typical deflection mode of shear wall–frame building.

Expressing the displacement in terms of the normal coordinates

$$x(z, t) = \sum_{i=1}^{N} \phi_i(z) q_i(t) \tag{3.40}$$

where ϕ_i is the i-th vibration mode shape and q_i is the i-th normal coordinate, and using the orthogonality conditions given in Chapter 1, Eq. (3.39) can be expressed as

$$m_i^* \ddot{q}_i + c_i^* \dot{q}_i + k_i^* q_i = p_i^*; \qquad i = 1 \text{ to } N \tag{3.41}$$

where m_i^*, c_i^* and k_i^* are the generalized mass, damping and stiffness in the i-th mode of vibration, while p_i^* is the generalized force. These are determined from

$$m_i^* = \int_0^H m(z)\phi_i^2(z)\,dz$$

$$c_i^* = \int_0^H c(z)\phi_i^2(z)\,dz$$

$$k_i^* = \int_0^H EI(z)\phi_i''''(z)\phi_i(z)\,dz - \int_0^H GA(z)\phi_i''(z)\phi_i(z)\,dz$$

$$p_i^* = \int_0^H f(z,t)\phi_i(z)\,dz$$

$$= \rho C_D B \int_0^H \bar{U}(z)\,u(z,t)\phi_i(z)\,dz \tag{3.42}$$

Equation (3.41) consists of a set of uncoupled equations, each representing a single-degree-of-freedom system. Thus the response in each normal coordinate can be obtained as in section 3.3.1:

$$S_{q_i}(n) = |H_i(n)|^2 S_{p_i^*}(n)\frac{1}{(k_i^*)^2} \tag{3.43}$$

where

$$|H_i(n)| = \frac{1}{\left(\left[1 - \left(\dfrac{n}{n_i}\right)^2\right]^2 + 4\zeta_i^2\left(\dfrac{n}{n_i}\right)^2\right)^{1/2}}$$

$$k_i^* = 4\pi^2 n_i^2 m_i^* \tag{3.44}$$

in which n_i and ζ_i are the frequency and damping ratio in the i-th mode. The spectral density of the generalized force takes the form

$$S_{p_i^*}(n) = \rho^2 C_D^2 B^2 \chi^2(n) \int_0^H\int_0^H \bar{U}(z_1)\bar{U}(z_2)S_{u_1 u_2}(r,n)\phi_i(z_1)\phi_i(z_2)\,dz_1\,dz_2 \tag{3.45}$$

where $S_{u_1 u_2}(r,n)$ is the cross spectral density defined in Eq. (3.14) with r being the distance between the coordinates z_1 and z_2. In Eq. (3.45), the aerodynamic admittance has been incorporated to account for the distortion caused by the structure to the turbulent velocity.

According to Eq. (3.15), Eq. (3.45) may be expressed as

$$S_{p_i^*}(n) = \rho^2 C_D^2 B^2 \chi^2(n) \int_0^H\int_0^H \phi_i(z_1)\phi_i(z_2)\bar{U}(z_1)\bar{U}(z_2)$$

$$\times \sqrt{S_{u_1}(n)}\sqrt{S_{u_2}(n)}\,\gamma(r,n)\,dz_1\,dz_2 \tag{3.46}$$

where $\gamma(r, n)$ is the square root of the coherence given in Eq. (3.15), and $S_u(n)$ is the spectral density of the turbulent velocity.

The variance of the i-th normal coordinate is obtained from

$$\sigma_{q_i}^2 = \int_0^\infty S_{q_i}(n) \, dn$$

$$= \frac{1}{(k_i^*)^2} \left[\int_0^{n_i - \Delta n} S_{p^*}(n) \, dn + \frac{\pi n_i}{4 \zeta_i} S_{p^*}(n_i) \right]$$

$$= \sigma_{Bq_i}^2 + \sigma_{Dq_i}^2 \tag{3.47}$$

in which σ_{Bq_i} and σ_{Dq_i} are the non-resonating and the resonating r.m.s. responses of the i-th normal coordinate. As the response due to various modes of vibration are statistically uncorrelated, the response of the system is given by

$$\sigma_x^2(z) = \sum_{i=1}^N \phi_i^2(z) \sigma_{Bq_i}^2 + \sum_{i=1}^N \phi_i^2(z) \sigma_{Dq_i}^2 \tag{3.48}$$

which gives the variance and hence the r.m.s. displacement at various heights.

It should be noted that in order to determine the total displacement t, the static deflection due to the mean drag load given in Eq. (3.25) must be included, which is determined conveniently as follows. The mean generalized force is given by

$$\bar{f}_i = \int_0^H \tfrac{1}{2} \rho C_D \bar{U}^2(z) B \phi_i(z) \, dz$$

$$= \tfrac{1}{2} \rho C_D B \int_0^H \bar{U}^2(z) \phi_i(z) \, dz \tag{3.49}$$

Then the mean displacement is determined from

$$\bar{x}(z) = \sum_{i=1}^N \phi_i(z) \left[\frac{\bar{f}_i}{(2 \pi n_i)^2 m_i^*} \right] \tag{3.50}$$

The r.m.s. acceleration is obtained from

$$\sigma_{\ddot{x}}(z) = \left[\sum_{i=1}^N (2 \pi n_i)^4 \phi_i^2(z) \sigma_{Dq_i}^2 \right]^{1/2} \tag{3.51}$$

The dynamic shear and bending moment at any height z are obtained from the vibratory inertia forces in each mode and then by summing the modal contributions. For example, the variance of the base shear, Q, is obtained as

$$\sigma_Q^2 = \sum_{i=1}^N \sigma_{Qi}^2 \tag{3.52}$$

where

$$\sigma_{Qi} = \sigma_{Dq_i} \int_0^H m(z)(2 \pi n_i)^2 \phi_i(z) \, dz \tag{3.53}$$

3.3.3 Evaluation of Peak Response

The probability of the response exceeding a certain magnitude is determined using a peak factor on the r.m.s. response. Devenport [3.16] recommended the following expression for 50% probability of exceedence:

$$g = \sqrt{[2\ln(\nu T_0)]} + \frac{0.577}{\sqrt{[2\ln(\nu T_0)]}} \tag{3.54}$$

where g is the peak factor, ν the expected frequency at which the fluctuating response crosses the zero axis with positive slope and T_0 the period (usually 3600 s) during which the peak response is assumed to occur.

For resonant response, ν is equal to the natural frequency and, thus, the peak factor for the resonant response g_D is obtained from Eq. (3.54) by setting $\nu = n$. For the non-resonating or broad-band response, the peak factor has been evaluated to be [3.17]:

$$g_B = 3.5$$

Using these peak factors, the most probable maximum value of the load effect, E, such as displacement, shear, bending moment, etc., is determined as follows:

$$E_{\max} = \bar{E} + [(g_B \sigma_{BE})^2 + (g_D \sigma_{DE})^2]^{1/2} \tag{3.55}$$

where σ_{BE} and σ_{DE} are the non-resonating and the resonating components of the load effect and $\bar{E}$ is the load effect due to the mean wind.

Example 3.1. An observation tower with dimensions $H = 70$ m, $B = 60$ m and $D = 12$ m (see Fig. 3.15) is situated in a suburban terrain. The period of the tower is 1.6 s. The damping ratio is estimated to be 1%. The mass concentrated at the top of the tower is 325 000 kg. Idealizing the tower as a point structure, determine the maximum drift and base shear for a 50 year wind. The reference wind speed at 10 m height is 15 m/s. Assume the drag coefficient C_D is 1.3 and the density of air is 1.2 kg/m^3.

Solution. From Table 3.1, for suburban terrain the roughness length z_0 is 0.3 and the height of the zero plane above the ground is 5 m. Thus, from Eq. (3.2), the friction velocity

$$u_* = \frac{k\bar{U}(z)}{\ln\left(\dfrac{z-d}{z_0}\right)} = \frac{0.4(15)}{\ln\left(\dfrac{10-5}{0.3}\right)} = 2.13 \text{ m/s}$$

$$\bar{U}(70) = \frac{1}{0.4} \times 2.13 \times \ln\left(\frac{70-5}{0.3}\right) = 28.64 \text{ m/s}$$

The power spectrum of the wind at 70 m height is determined as follows. The reduced frequency

$$f = \frac{nz}{\bar{U}(z)} = \frac{n(70)}{28.64} = 2.44n$$

From Eq. (3.13):

$$S_u(H, n) = \frac{u_*^2}{n} \frac{200f}{(1 + 50f)^{5/3}} = \frac{2.13^2 \times 200 \times 2.44}{(1 + 50 \times 2.44)^{5/3}}$$

$$= \frac{2214}{(1 + 122n)^{5/3}}$$

Response to mean wind
The mean force

$$\bar{F} = \tfrac{1}{2}\rho C_D BD[\bar{U}(70)]^2$$

$$= \tfrac{1}{2}(1.2)(1.3)(6)(12)(28.64^2)$$

$$= 46\,065 \text{ N}$$

The stiffness of the tower

$$K = 325\,000 \times \left(\frac{2\pi}{1.6}\right)^2$$

$$= 5 \times 10^6 \text{ N/m}$$

Thus the mean displacement

$$\bar{X}_{\text{mean}} = \frac{46\,065}{5 \times 10^6} \times 10^3 = 9.21 \text{ mm}$$

Non-resonant displacement
The variance of the non-resonant displacement is given by Eqs. (3.37) and (3.33) as

$$\sigma_B^2 = \frac{1}{K^2} \rho^2 \bar{U}^2(H)B^2 D^2 C_D^2 \int_0^{n_1 - \Delta n} \chi^2(n)S_u(n) \, dn$$

The admittance function $\chi(n)$ can be taken to be unity for larger eddies with low frequencies, which are responsible for the non-resonant displacement.
Furthermore, from Eq. (3.6):

$$\sigma_u^2 = \beta u_*^2 = \int_0^\infty S_u(n) \, dn$$

where $\beta = 5.25$ from Table 3.2. Thus

$$\sigma_B^2 = \frac{1.2^2 \times (28.64)^2 \times 6^2 \times 12^2 \times 1.3^2 \times 5.25 \times 2.13^2}{(5 \times 10^6)^2} \, \text{m}^2$$

$$= 3.14 \, \text{mm}$$

Resonant displacement

The admittance function at $n_1 = \dfrac{1}{1.6} = 0.625$ Hz

$$\chi(n_1) = \frac{1}{1 + \left[\dfrac{2n_1\sqrt{A}}{\bar{U}(H)}\right]^{4/3}}$$

$$= \frac{1}{1 + \left[\dfrac{2 \times 0.625\sqrt{12 \times 6}}{28.64}\right]^{4/3}} = 0.79$$

From Eq. (3.33):

$$S_F(n_1) = \rho^2 \bar{U}^2(H)B^2D^2C_D^2 \, \chi^2(n_1)S_u(H, n_1)$$

$$= 1.2^2 \times 28.64^2 \times 6^2 \times 12^2 \times 1.3^2 \times (0.79)^2$$

$$\times \left[\frac{2214}{(1 + 122 \times 0.625)^{5/3}}\right]$$

$$= 10.2 \times 10^6 \, \text{N}^2/\text{Hz}$$

From Eq. (3.37), the variance of resonant displacement

$$\sigma_D^2 = \frac{\pi n_1 S_F(n_1)}{4\zeta_1 K^2}$$

$$= \frac{\pi(0.625)(10.2 \times 10^6)}{4(0.01)(5 \times 10^6)^2} \, \text{m}^2$$

$$\sigma_D = 4.48 \, \text{mm}$$

The r.m.s. acceleration

$$\ddot{\sigma}_D = (2\pi \times 0.625)^2 \frac{4.48}{10^3} = 0.069 \, \text{m/s}^2$$

From Eq. (3.54), the peak factor for the resonant response,

$$g_D = \sqrt{2 \ln(0.625 \times 3600)} + \frac{0.577}{\sqrt{2 \ln(0.625 \times 3600)}} = 4.08$$

Assuming a peak factor of 3.5 for a non-resonant response, the most probable maximum displacement is

$$= 9.21 + \sqrt{(3.5 \times 3.14)^2 + (4.08 \times 4.48)^2} \text{ mm}$$

$$= 30.54 \text{ mm}$$

$$\text{Maximum drift} = \frac{30.54}{70\,000} = \frac{1}{2292}$$

Maximum dynamic base shear $= 325\,000 \times 4.08 \times 0.069 = 91\,494 \text{ N}$

Maximum base shear due to non-resonant displacement

$$= 5 \times 10^6 \times \frac{3.5}{10^3} \times 3.14 = 54\,950 \text{ N}$$

Mean base shear $= 5 \times 10^6 \times \frac{9.21}{10^3} = 46\,050 \text{ N}$

Thus, the most probable maximum base shear

$$= 46\,050 + \sqrt{(54\,950)^2 + (91\,494)^2} = 152\,777 \text{ N}$$

Example 3.2. A rectangular building of height $H = 194$ m is situated in a suburban terrain. The breadth B and width D of the building are 56 m and 32 m respectively. The period of the building corresponding to the fundamental sway mode is 5.15 s. The values of the mode shape at various heights are:

H (m)	0	20	40	75	95	135	150	170	194
ϕ	0	0.032	0.096	0.248	0.365	0.611	0.746	0.849	1.0

The generalized mass and damping ratio corresponding to this mode are 18×10^6 kg and 2% respectively.

Assuming that the mean wind profile follows the power law with a power law coefficient $\alpha = 0.22$, determine the maximum drift for a 50 year wind storm of 21 m/s at 10 m height, blowing normal to the breadth of the building. The friction velocity is 2.96 m/s, the drag coefficient C_D is 1.3 and the density of air ρ is 1.2 kg/m^3.

Solution. The mean height of the building $\bar{H} = 97$ m

$$\bar{U}(97) = \bar{U}(10)\left(\frac{97}{10}\right)^{0.22}$$

$$= 21\left(\frac{97}{10}\right)^{0.22} = 34.6 \text{ m/s}$$

At mid-height, the reduced frequency

$$f = \frac{n\bar{H}}{\bar{U}(\bar{H})} = \frac{97n}{34.6} = 2.8n$$

From Eq. (3.13), the spectrum of turbulent wind is given by

$$S_u(\bar{H}, n) = \frac{2.96^2 \times 200 \times 2.8}{(1 + 50 \times 2.8n)^{5/3}} = \frac{4906}{(1 + 140n)^{5/3}}$$

Response to mean wind
From Eq. (3.49), the mean generalized force

$$\bar{f}_1 = \tfrac{1}{2}\rho C_D B \int_0^H \bar{U}^2(z)\phi_1(z)\,dz$$

$$= \tfrac{1}{2}\rho C_D B[\bar{U}(\bar{H})]^2 \left(\frac{1}{\bar{H}}\right)^{2\alpha} \int_0^H z^{2\alpha}\phi_1(z)\,dz$$

$$= \tfrac{1}{2} \times 1.2 \times 1.3 \times 56 \times (34.6)^2 \left(\frac{1}{97}\right)^{0.44} \times 684$$

$$= 4.8 \times 10^6 \text{ N}$$

The generalized stiffness

$$k_1^* = \left(\frac{2\pi}{5.15}\right)^2 \times 18 \times 10^6$$

$$= 26.8 \times 10^6 \text{ N/m}$$

Thus the mean displacement

$$\bar{X} = \frac{4.8 \times 10^6}{26.8 \times 10^6} \times 10^3 = 179 \text{ mm}$$

Resonant displacement
The resonant frequency

$$n_1 = \frac{1}{5.15} = 0.194 \text{ Hz}$$

and therefore

$$S_u(\bar{H}, n_1) = \frac{4906}{(1 + 140 \times 0.194)^{5/3}} = 18.8 \text{ m}^2/\text{s}$$

The admittance function, from Eq. (3.32), becomes

$$\chi(n) = \cfrac{1}{1 + \left[\cfrac{2n\sqrt{56 \times 194}}{34.6}\right]^{4/3}}$$

$$= \frac{1}{1 + 10.96n^{4/3}}$$

$$\chi(n_1) = 0.45$$

From Eq. (3.46):

$$S_{p\dot{p}}(n) = \rho^2 C_D^2 B^2 \chi^2(n) S_u(\bar{H}, n) \frac{[\bar{U}(\bar{H})]^2}{\bar{H}^{2\alpha}}$$

$$\times \int_0^H \int_0^H \phi_1(z_1)\phi_1(z_2)z_1^\alpha z_2^\alpha \gamma(z_1, z_2, n) \, dz_1 \, dz_2$$

The square root of coherence γ is determined from Eq. (3.16), considering only the vertical correlation. Thus

$$S_{p\dot{p}}(n_1) = 1.2^2 \times 1.3^2 \times 56^2 \times (0.45)^2 \times 18.8 \times \frac{(34.6)^2}{97^{0.44}} \times 16\,900$$

$$= 7.85 \times 10^{10} \text{ N}^2/\text{Hz}$$

From Eqs. (3.47) and (3.48), the variance of the resonant displacement at the top of the building is obtained as

$$\sigma_D^2 = \phi_1^2(H)\sigma_{Dq_1}^2 = \frac{1}{(k_1^*)^2} \frac{\pi n_1}{4\zeta_1} S_{p\dot{p}}(n_1)$$

$$= \left(\frac{1}{26.8 \times 10^6}\right)^2 \left(\frac{\pi(0.194)}{4(0.02)}\right)(7.85 \times 10^{10})10^6 \text{ mm}^2$$

$$\sigma_D = 28.9 \text{ mm}$$

Non-resonant displacement
The variance of the non-resonant displacement at the top of the building is determined from Eqs. (3.47) and (3.48) as

$$\sigma_B^2 = \phi_1^2(H)\sigma_{Bq_1}^2 = \frac{1}{(k_1^*)^2} \int_0^{n_1 - \Delta n} S_{p\dot{p}}(n) \, dn$$

$$= \frac{565 \times 10^9}{(26.8 \times 10^6)^2} \times 10^6 \text{ mm}$$

$$\sigma_B = 28 \text{ mm}$$

The peak factor g_D for the resonant response is determined from Eq. (3.54) as 3.78. Using a peak factor of 3.5 for the non-resonant response, the most probable maximum displacement is

$$X_{max} = \bar{X} + \sqrt{(g_B\sigma_B)^2 + (g_D\sigma_D)^2}$$

$$= 179 + \sqrt{(3.5 \times 28)^2 + (3.78 \times 28.9)^2}$$

$$= 326 \text{ mm}$$

The most probable maximum drift would be

$$\frac{326}{194\,000} = \frac{1}{595}$$

3.4 Across-wind Response

For most modern tall buildings, the across-wind response is more significant than the along-wind response. Across-wind vibration of structures is caused by the combination of forces from three sources: (1) buffeting by the turbulence in the across-wind direction, (2) wake excitation due to vortex shedding and (3) aeroelastic phenomena such as lock-in, galloping and flutter.

The across-wind force due to lateral turbulence in the approaching flow is generally small compared with the effects due to other mechanisms. Lock-in, galloping and flutter are displacement-dependent excitations and in practice tall buildings are not prone to galloping or flutter. Galloping can be significant for flexible, lightly damped and slender tower-like structures [3.18, 3.19], whereas flutter is most likely to occur in bridge decks [3.20] or cantilevered roofs. Figure 3.18 illustrates schematically the range of reduced frequencies over which various sources of across-wind excitation prevail.

Lock-in is the term used to describe large-amplitude across-wind motion which occurs when the vortex shedding frequency is close to the natural frequency. If the across-wind response exceeds a certain critical value, it causes an increase in the excitation force, which in turn increases the response. The vortex shedding frequency tends to couple with the natural frequency of the structure for a range of wind velocities, and the large-amplitude response will persist. Lock-in is likely to occur only in the case of structures with relatively low stiffness and low damping, operating near the critical wind velocity given by

$$\bar{U}_{crit} = \frac{n_0 B}{S} \tag{3.56}$$

where $\bar{U}_{crit}$ is the critical wind speed, B the breadth of the structure normal to the wind stream, n_0 (Hz) the fundamental natural frequency of the structure in the across-wind direction and S the Strouhal number.

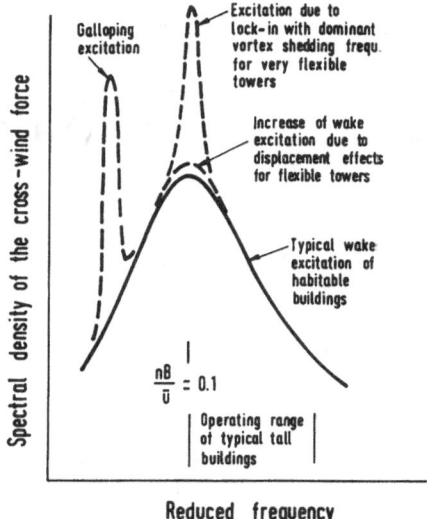

Fig. 3.18. Spectra of across-wind forces. (Reproduced by permission of Cambridge University Press [3.24].)

In practice, chimneys and stacks are the only structures commonly affected by lock-in. Structures should be designed so that lock-in effects do not occur during their anticipated life. If the r.m.s. displacement at the top of the structure is less than a certain critical value, then lock-in will not occur. For square tall buildings, the critical r.m.s. displacements σ_{yc} expressed as a ratio with respect to the breadth (σ_{yc}/B) are [3.21] approximately 0.015, 0.025 and 0.045, respectively, for open terrain, suburban terrain and city centres. For circular sections with diameter D, the value of σ_{yc}/D is approximately 0.006 for suburban terrain.

Thus, for buildings the most common cause for across-wind motion is wake excitation. Turbulence in the atmospheric boundary layer affects the regularity of vortex shedding. However, the shed vortices have a predominant period which can be determined from an appropriate Strouhal number. Because vortex shedding is random, the fluctuating across-wind force is effectively broad band, as shown in Fig. 3.19. The bandwidth and the energy concentration near the vortex shedding frequency depend on the geometry of the building and the characteristics of the approach flow.

The response due to this across-wind random excitation can be determined using random vibration theory. Idealizing the tall building as a line-like structure, as in section 3.3.2, the across-wind displacement $y(z, t)$ can be expressed in terms of the normal coordinates $r_i(t)$ as

$$y(z, t) = \sum_{i=1}^{N} \psi_i(z) r_i(t) \tag{3.57}$$

where $\psi_i(z)$ is the i-th vibration mode in the across-wind direction and N is

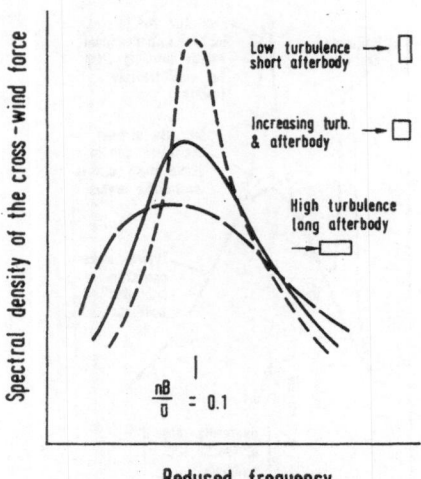

Fig. 3.19. Effects of turbulence intensity and afterbody length on across-wind force spectra.

the total number of modes considered to be significant. The governing equation of motion in terms of generalized mass m_i^*, generalized damping c_i^* and generalized stiffness k_i^* takes the form

$$m_i^* \ddot{r}_i + c_i^* \dot{r}_i + k_i^* r_i = f_i^*(t) \qquad i = 1 \text{ to } N \tag{3.58}$$

in which

$$m_i^* = \int_0^H m(z)\psi_i^2(z)\,\mathrm{d}z$$

$$k_i^* = (2\pi n_i)^2 m_i^*$$

$$c_i^* = 2\zeta_i \sqrt{m_i^* k_i^*}$$

$$f_i^*(t) = \int_0^H f(z, t)\psi_i(z)\,\mathrm{d}z \tag{3.59}$$

where H is the height of the building, $m(z)$ the mass per unit length, n_i the frequency of the i-th mode in the across-wind direction, ζ_i the damping ratio in the i-th mode, $f(z, t)$ the across-wind force per unit height and $f_i^*(t)$ the generalized across-wind force in the i-th mode. The spectral density of each normal coordinate can be determined from

$$S_{r_i}(n) = \frac{|H_i(n)|^2}{(k_i^*)^2} S_{f_i^*}(n) \tag{3.60}$$

where $|H_i(n)|$ is the mechanical admittance function and $S_{f_i^*}(n)$ the power spectral density of the generalized across-wind force.

Unlike the case of drag force due to along-wind buffeting, it is not practical to relate analytically the velocity fluctuation in the approach flow to the pressure acting on the sides of the building in a separated flow. Thus across-wind spectra are determined experimentally. Kareem [3.22] has proposed the following empirical expression for the spectral density of across-wind force S_f for square buildings:

$$\frac{nS_f(z, n)}{\sigma_f^2} = \alpha^*\beta^*\left(\frac{n}{n_s}\right)^{0.9} \qquad n \leq n_s$$

$$= \alpha^*\beta^*\left(\frac{n}{n_s}\right)^{0.3} \qquad n \geq n_s \tag{3.61}$$

where

$$\alpha^* = \frac{b}{\left[1 - \left(\frac{n}{n_s}\right)^2\right]^2 + \left[2b\left(\frac{n}{n_s}\right)\right]^2}$$

$$\beta^* = 1.32\left[\left(\frac{1}{3\alpha}\right)^{1/2} + 0.154\left(1 - \frac{z}{H}\right)^{3.5}\right] \tag{3.62}$$

where n_s is the shedding frequency $= S\bar{U}(z)/B$, S the Strouhal number, $\bar{U}(z)$ the mean speed at height z, B the breadth of the building, σ_f^2 the mean square value of the fluctuating across-wind force, α the exponent term in the power law, b the bandwidth coefficient $= \sqrt{2}I(z)$ and $I(z)$ the turbulence intensity at height z.

According to Eq. (3.59), the normalized power spectrum for the generalized across-wind force in the i-th mode may be expressed as

$$\frac{nS_{f_i^*}(n)}{[\frac{1}{2}\rho\bar{U}^2(H)BH]^2} = \frac{1}{H^2}\int_0^H\int_0^H\left(\frac{nS_f(z_1, n)}{\sigma_f^2(z_1)}\right)^{1/2}\left(\frac{nS_f(z_2, n)}{\sigma_f^2(z_2)}\right)^{1/2}$$

$$\times C_L(z_1)C_L(z_2)\,\text{Coh}(z_1, z_2, n)\psi_i(z_1)\psi_i(z_2)\,dz_1\,dz_2 \tag{3.63}$$

in which

$$C_L(z) = \frac{\sigma_f(z)}{\frac{1}{2}\rho B\bar{U}^2(H)}$$

$$\text{Coh}\left(\frac{\Delta z}{H}, n\right) = \exp\left\{\left(-\frac{\Delta z}{H}\right)^{1/3}\left[1 - \frac{\Delta z}{H}\right]^{1/2}\left(\frac{nB}{\bar{U}(H)}\right)\right\}$$

$$\times\left\{\exp\left(\frac{2\Delta z}{H}\right)\alpha(0.88 + \alpha)^2 + \frac{2}{5}\left[1 - \exp\left(\frac{\Delta z}{H}\right)\right]\right\}$$

$$\times\left\{1 + 5\left(\frac{1}{3} - \alpha\right)\left(1 - \exp\left[-\left(\frac{\Delta z}{H}\right)^2\right]\right)\right\}$$

$$\times \left[\text{Cos}\left\{ \left(\frac{20\pi}{1-\pi} \left(\frac{nB}{\bar{U}(H)} \right) - 1 \right\} \right]; \qquad \frac{nB}{\bar{U}(H)} \leq F^*$$

$$= \text{Coh}\left(\frac{\Delta z}{H}, F^* \right) \exp\left(-20\left(\frac{\Delta z}{H} \right)^{1.2} \left(\frac{nB}{\bar{U}(H)} \right)^{1/2} \right); \qquad \frac{nB}{\bar{U}(H)} > F^*$$

$$(3.64)$$

where $\Delta z = |z_1 - z_2|$ and $F^* = 1.25(1 - \alpha)/10$. Equation (3.63) can be used to generate force spectra for different approach flow characteristics and building heights. Using linear mode, the generalized force spectra obtained from Eq. (3.63) for open country, suburban and urban flow conditions are presented in Fig. 3.20 for an aspect ratio of $1:6$ [3.22].

The variance of the normal coordinate is given by

$$\sigma_{r_i}^2 = \int_0^\infty \frac{|H_i(n)|^2}{(k_i^*)^2} S_{f_i^*}(n) \, dn \tag{3.65}$$

Hence, the variance of the across-wind displacement is obtained from

$$\sigma_y^2(z) = \sum_{i=1}^N \psi_i^2(z)\sigma_{r_i}^2 \tag{3.66}$$

In Eq. (3.65), if the contribution from the non-resonating component is neglected, then the r.m.s. response of the across-wind displacement is determined from

$$\sigma_y(z) = \left[\sum_{i=1}^N \frac{\psi_i^2(z)}{(2\pi n_i)^4 (m_i^*)^2} \left(\frac{\pi n_i}{4\zeta_i} \right) S_{f_i^*}(n_i) \right]^{1/2} \tag{3.67}$$

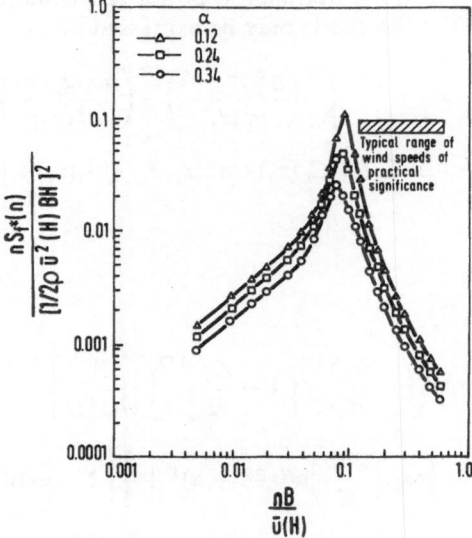

Fig. 3.20. Generalized force spectra for open country, suburban and urban flow conditions. (Reproduced by permission of the publishers Butterworth Heinemann Ltd. [3.22].)

For convenient use of the above equation, the generalized force spectra obtained experimentally by Kwok and Melbourne [3.23] and Saunders and Melbourne [3.24] are presented in Fig. 3.21 for various aspect ratios of square and rectangular buildings deflecting in a linear mode.

Vikery [3.25] has proposed the following empirical expression for the across-wind displacement of various cross-sections, as shown in Fig. 3.22, with

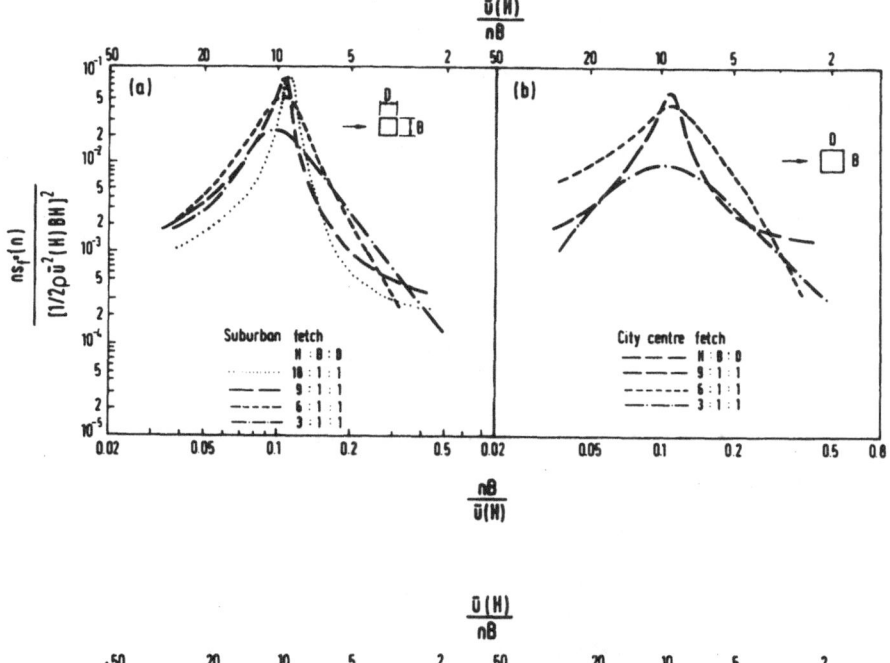

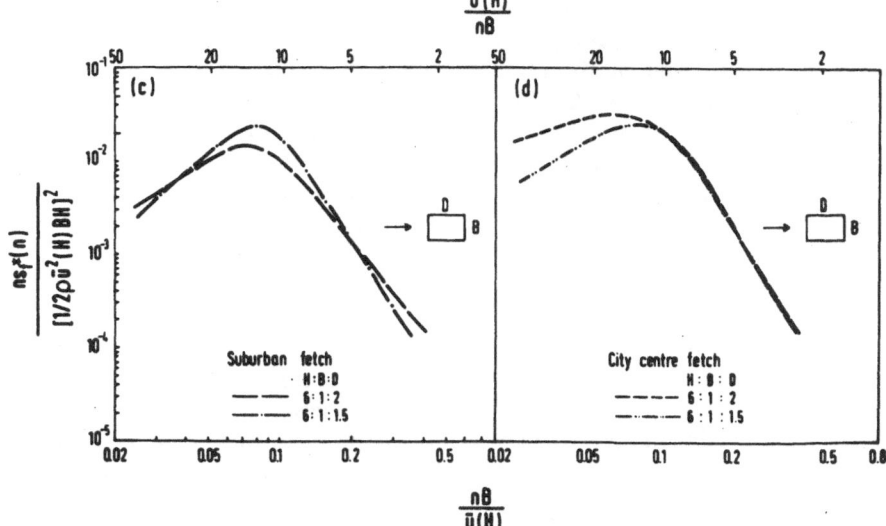

Fig. 3.21. Generalized force spectra for square and rectangular buildings in suburban and city centre fetches.

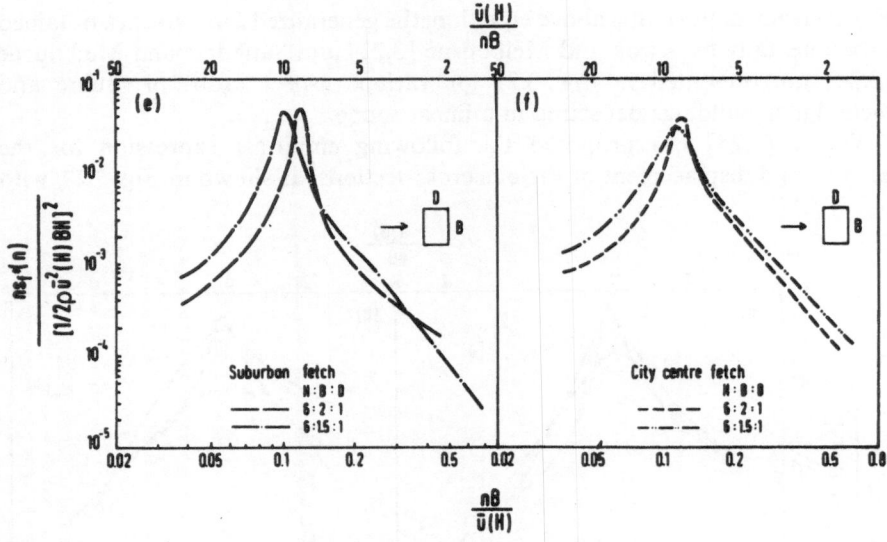

Fig. 3.21. (Continued).

an average mass density of 200 kg/m³ and a damping ratio of 1%:

$$\frac{\sigma_y(H)}{\sqrt{A}} = c \left[\frac{\bar{U}(H)}{n_1 \sqrt{A}} \right]^{3.5} \frac{1}{\sqrt{\zeta_1}} \frac{\rho}{\rho_b} \tag{3.68}$$

where $\sigma_y(H)$ is the r.m.s. tip displacement, H the height of the building, A the cross-sectional area of the building, $\bar{U}(H)$ the mean wind speed at the tip of the building, n_1 the fundamental frequency of vibration, ζ_1 the damping ratio, ρ the air density and ρ_b the mass density of the building. The constant c is determined empirically as 0.00015 ± 0.00006. The peak displacement is obtained by multiplying the r.m.s. value by a peak factor of 4. In view of the empirical nature of Eq. (3.68), it would be unwise to apply this equation to buildings with properties well removed from the range investigated.

The r.m.s. acceleration at the top of the structure can be estimated as

$$\sigma_y(H) = (2\pi n_1)^2 \sigma_y(H) \tag{3.69}$$

Example 3.3. Consider the building of Example 3.2. If the period of vibration in the across-wind direction is 4.6 s, assuming a linear mode determine the acceleration in the across-wind direction. The generalized mass corresponding to the linear mode is 17.5×10^6 kg and the damping in this mode of oscillation is 2%.

Solution. The building is rectangular with an aspect ratio of

$$H : B : D = 6 : 1.75 : 1$$

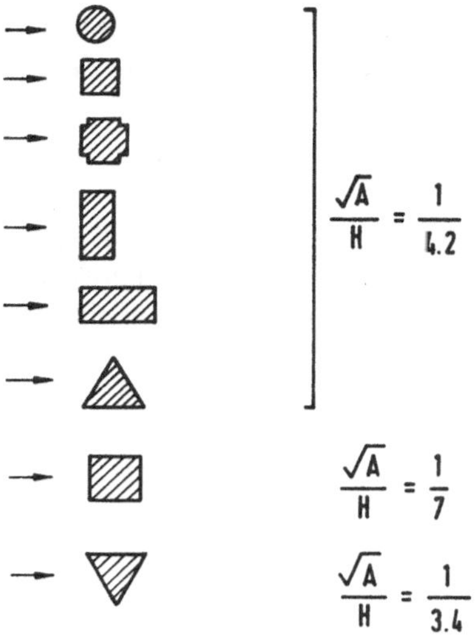

$$\frac{\sqrt{A}}{H} = \frac{1}{4.2}$$

$$\frac{\sqrt{A}}{H} = \frac{1}{7}$$

$$\frac{\sqrt{A}}{H} = \frac{1}{3.4}$$

Fig. 3.22. Characteristics of models tested for Eq. (3.68).

Since the building is in a suburban terrain, the generalized across-wind force can be determined from Fig. 3.21e. The wind speed at the tip of the building

$$\bar{U}(H) = \bar{U}(10) \times \left(\frac{194}{10}\right)^{0.22} = 40.3 \text{ m/s}$$

The reduced frequency

$$\frac{n_1 B}{\bar{U}(H)} = \frac{0.217 \times 56}{40.3} = 0.3$$

so from Fig. 3.21e

$$S_{f_1^*}(n_1) = \left(\frac{0.0004}{0.217}\right)(\tfrac{1}{2} \times 1.2 \times 40.3^2 \times 56 \times 194)^2$$

$$= 2.6 \times 10^{11} \text{ N}^2/\text{Hz}$$

From Eq. (3.67):

$$\sigma_y(H) = \left[\left(\frac{\pi \times 0.217}{4 \times 0.02}\right)(2.6 \times 10^{11}) \frac{1}{(2\pi \times 0.217)^4 (17.5 \times 10^6)^2}\right]^{1/2}$$

$$= 0.046 \text{ m}$$

Assuming a peak factor of 4, the peak acceleration in the across-wind direction

$$\ddot{\sigma}_y(H) = 4 \times 0.046 \times (2\pi)^2 (0.217)^2$$

$$= 0.34 \text{ m/s}^2 \ (3.4\% \ g)$$

Alternatively, the r.m.s. across-wind displacement may be evaluated from Eq. (3.68), using the average density of the building determined as

$$\rho_b = \frac{3m_1^*}{AH} = \frac{3 \times 17.5 \times 10^6}{56 \times 32 \times 194} = 151 \text{ kg/m}^3$$

In Eq. (3.68), the coefficient c ranges from 0.00009 to 0.00021. Correspondingly, the r.m.s. displacement at the tip would lie between 0.038 m and 0.088 m. It should be noted that the value obtained from Fig. 3.21 lies within the range predicted by Eq. (3.68).

3.5 Torsional Response

Buildings will be subjected to torsional motion when the instantaneous point of application of the resultant aerodynamic load does not coincide with the centre of mass and/or the elastic centre. Even in a symmetrical building, along-wind forces can cause torsional motion as a result of uncorrelated wind loads acting across the breadth of the building. However, the major sources of dynamic torque are the flow-induced asymmetries in the lift force and the pressure fluctuation on the leeward side caused by vortex shedding. Any eccentricities between the centre of mass and the centre of stiffness present in aysmmetrical buildings can amplify the torsional effects.

Safak and Foutch [3.26] have presented a method for estimating the along-wind, across-wind and torsional responses of rectangular buildings in the frequency domain. Recently Balendra et al. [3.27] have presented a time-domain approach to estimate the coupled lateral–torsional motion of buildings due to along-wind turbulence and across-wind forces, and torque due to wake excitation. The experimentally measured power spectra of across-wind forces and torsional moments [3.28] were used in this analysis. These methods are useful at the final stages of design, since specific details which are unique for a particular building can be easily incorporated into the analytical model. A useful method to assess the torsional effects at the preliminary design stage is given by the following empirical relation [3.7] which yields the peak base torque induced by wind speed $\bar{U}(H)$ at the top of the building as

$$T_{\text{peak}} = \Psi\{\bar{T} + g_T T_{\text{rms}}\} \tag{3.70}$$

where Ψ is a reduction coefficient, g_T the torsional peak factor equal to 3.8, and $\bar{T}$ and T_{rms} the mean and the root mean square base torques which are given by

$$\bar{T} = 0.038\rho L^4 H n_T^2 U_r^2$$

$$T_{\text{rms}} = 0.00167 \frac{1}{\sqrt{\zeta_T}} \rho L^4 H n_T^2 U_r^{2.68} \tag{3.71}$$

in which

$$U_r = \frac{\bar{U}(H)}{n_T L}$$

$$L = \frac{\int |r|\, ds}{\sqrt{A}} \tag{3.72}$$

where ρ is the density, H the height of the building, n_T and ζ_T the frequency and damping ratio in the fundamental torsional mode of vibration, $|r|$ the distance between the elastic centre and the normal to an element ds on the boundary of the building and A the cross-sectional area of the building. The expressions for $\bar{T}$ and T_{rms} are obtained for the most unfavourable directions for the mean and r.m.s. values of the base torque. In general these directions do not coincide, and furthermore will not be along the direction of the extreme winds expected to occur at the site. As such, a reduction coefficient Ψ $(0.75 < \Psi \leq 1)$ is incorporated in Eq. (3.70).

Assuming a linear fundamental mode shape, the peak-torsional-induced horizontal acceleration at the top of the building at a distance a from the elastic centre is given by [3.29]

$$a\ddot{\theta} = \frac{2ag_T T_{rms}}{\rho_b BDH r_m^2} \tag{3.73}$$

where $\ddot{\theta}$ is the peak angular acceleration, ρ_b the mass density of the building, B and D the breadth and depth of the building and r_m the radius of gyration. For a rectangular building with uniform mass density

$$r_m^2 = \tfrac{1}{12}(B^2 + D^2) \tag{3.74}$$

Example 3.4. If the torsional frequency of the building in Example 3.3 is 0.5 Hz, assuming a linear mode and 2% damping ratio, determine the peak acceleration at the corner of the building due to torsional motion. Take the centre of rigidity to be at the geometric centre of the building.

Solution. For a rectangular building

$$\int |r|\, ds = \tfrac{1}{2}(B^2 + D^2)$$

Thus from Eq. (3.72):

$$L = \frac{1}{\sqrt{BD}}(B^2 + D^2)^{\frac{1}{2}} = 49.1 \text{ m}$$

$$U_r = \frac{\bar{U}(H)}{n_T L} = \frac{40.3}{0.5 \times 49.1} = 1.64$$

From Eq. (3.71):

$$T_{rms} = 0.00167\left(\frac{1}{\sqrt{0.02}}\right)(1.2)(49.1)^4(194)(0.5)^2(1.64)^{2.68}$$

$$= 15 \times 10^6 \, \text{N m}$$

The corner of the building is at a distance $a = 32.2$ m.

Thus from Eq. (3.73), the peak torsional acceleration of the corner

$$a\ddot{\theta} = \frac{2 \times 32.2 \times 3.8 \times 15 \times 10^6}{151 \times 56 \times 32 \times 194 \times 346.7} = 0.202 \, \text{m/s}^2$$

The above acceleration should be added vectorially with the peak along-wind and peak across-wind accelerations obtained in Examples 3.2 and 3.3, respectively. A reduction factor 0.80 may be employed while summing the peak responses, as the individual peaks may not occur simultaneously.

3.6 Serviceability Requirements

With the development of light-weight high-strength materials, the recent trend is to build tall and slender buildings. The design of such buildings is often governed by the need to limit wind-induced accelerations and drifts to acceptable levels for human comfort and integrity of non-structural components, respectively. Thus to check for serviceability of tall buildings, the peak resultant horizontal acceleration and displacement due to the combination of along-wind, across-wind and torsional loads are required. As an approximate estimation, the peak effects due to along-wind, across-wind and torsional responses may be determined individually and combined vectorially. A reduction factor of 0.8 may be used on the combined value to account for the fact that, in general, the individual peaks do not occur simultaneously. If the calculated combined effect is less than any of the individual effects, then the latter should be considered for the design.

British Standard BS 6611, 1984 [3.30] defines the comfort criterion as complaint by more than 2% of people in the upper floors of the building during the worst 10 minutes of a storm with a return period of 1 in 5 years. This is shown in Fig. 3.23 in terms of the r.m.s. acceleration for different frequencies [3.31]. However, as a rule of thumb the allowable peak acceleration is taken to be 20 milli g (0.2 m/s^2).

The allowable drift, defined as the resultant peak displacement at the top of the building divided by the height of the building, is generally taken to be around 1/500.

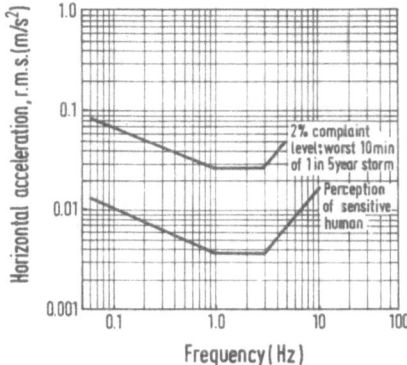

Fig. 3.23. Human response to horizontal motion.

References

3.1. *American National Standard Building Code Requirements for Minimum Design Loads in Buildings and Other Structures*, A 58.1, American National Standards Institute, New York, 1982.

3.2. von Kármán T, Turbulence and skin friction, *Journal of Aeronautical Sciences* 1934; 1.

3.3. *Characteristics of Atmospheric Turbulence Near the Ground, Part II: Single Point Data for Strong Winds (Neutral Atmosphere)*, Item 85020, Engineering Sciences Data Unit, London, 1985.

3.4. Biétry J, Sacré C and Simiu E, Mean wind profiles and changes of terrain roughness, *Journal of Structural Division, ASCE* 1978; 104: 1585–1593.

3.5. Counihan J, Adiabatic atmospheric boundary layers: a review and analysis of data from the period 1880–1972, *Atmospheric Environment* 1975; 9: 871–905.

3.6. Duchêne-Marullaz P, Effects of high roughness on the characteristics of turbulence in the case of strong winds, *Proceedings of the Fifth International Conference on Wind Engineering*, Vol. 1, Pergamon Press, Oxford, 1980.

3.7. Simiu E and Scanlan R H, *Wind Effects on Structures*, 2nd edn, Wiley, New York, 1986.

3.8. Simiu E, Wind spectra and dynamic along wind response, *Journal of Structural Division, ASCE* 1974; 100: 1897–1910.

3.9. Devenport A G, The dependence of wind load upon meteorological parameters, *Proceedings of the International Research Seminar on Wind Effects on Buildings and Structures*, University of Toronto Press, Toronto, 1968, pp 19–82.

3.10. Vickery B J, On the reliability of gust loading factors, *Proceedings of the Technical Meeting Concerning Wind Loads on Buildings and Structures*, Building Science Series 30, National Bureau of Standards, Washington D.C., 1970, pp. 93–104.

3.11. *Wind Loading and Wind Induced Structural Response*, State-of-the-Art Report, Committee on Wind Effects, American Society of Civil Engineers, New York, 1987.

3.12. Devenport A G, The application of statistical concepts to the wind loading of structures, *Proceedings of the Institute of Civil Engineers* 1961; 19: 449–472.

3.13. Vickery B J, *On the Flow Behind a Coarse Grid and its Use as a Model of Atmospheric Turbulence in Studies Related to Wind Loads on Buildings*, Nat. Phys. Lab. Aero. Report 1143, 1965.

3.14. Robson J D, *An Introduction to Random Vibration*, Edinburgh University Press, 1963.

3.15. Heidebrecht A C and Smith B S, Approximate analysis of tall wall-frame structures, *Journal of Structural Division, ASCE* 1973; 99: 199–221.

3.16. Devenport A G, Note on the distribution of the largest value of a random function with application to gust loading, *Proceedings of the Institute of Civil Engineers* 1964; 28: 187–196.

3.17. *The Response of Flexible Structures to Atmospheric Turbulence*, Item 76001, Engineering Sciences Data Unit, London, 1976.

3.18. Novak M and Devenport A G, Aeroelastic instability of prisms in turbulent flow, *Journal of Engineering Mechanics Division, ASCE* 1970; 96: 17–39.

3.19. Kwok K C S and Melbourne W H, Freestream turbulence effects on galloping, *Journal of Engineering Mechanics Division, ASCE* 1980; 106: 273–288.

3.20. Scanlan R H and Tomko J J, Aerofoil and bridge deck flutter derivatives, *Journal of Engineering Mechanics Division, ASCE* 1971; 97: 1717–1737.

3.21. Rosati P A, *An Experimental Study of the Response of a Square Prism to Wind Load*, Faculty of Graduate Studies, BLWT II–68, University of Western Ontario, London, Ontario, Canada, 1968.

3.22. Kareem A, Model for predicting the across-wind response of buildings, *Journal of Engineering Structures* 1984; 6: 136–141.

3.23. Kwok K C S and Melbourne W H, Wind induced lock-in excitation of tall structures, *Journal of Engineering Mechanics Division, ASCE* 1981; 107: 57–72.

3.24. Saunders J W and Melbourne W H, Tall rectangular building response to cross-wind excitation, *Proceedings of the 4th International Conference on Wind Effects on Building Structures*, Cambridge University Press, 1975, pp 369–379.

3.25. Vickery B J, *Notes on Wind Forces on Tall Buildings*, Annex to Australian Standard 1170, Part 2 – 1973, SAA Loading Code Part 2 – Wind Force, Standards Association of Australia, Sydney, 1973.

3.26. Safak E and Foutch D A, Coupled vibrations of rectangular buildings subjected to normally incident random wind loads, *Journal of Wind Engineering and Industrial Aerodynamics* 1987; 26: 129–148.

3.27. Balendra T, Nathan G K and Kang K H, Deterministic model for wind induced oscillations of buildings, *Journal of Engineering Mechanics, ASCE* 1989; 115: 179–199.

3.28. Reinhold T A, Measurements of simultaneous fluctuating loads at multiple levels on a model of tall building in a simulated urban boundary layer, PhD thesis, Department of Civil Engineering, Virginia Polytechnic Institute and State University, 1977.

3.29. Greig L, Toward an estimate of wind induced dynamic torque on tall buildings, MSc thesis, Department of Engineering, University of Western Ontario, London, Ontario, 1980.

3.30. *BS 6611 Guide to Evaluation of the Response of Occupants of Fixed Structures, Especially Buildings and Offshore Structures, to Low Frequency Horizontal Motion (0.063 Hz to 1.0 Hz)*, British Standards Institution, London, 1984.

3.31. Irwin A W, Human response to dynamic motion of structures, *Structural Engineer* 1978; 56A: 237–244.

Chapter 4

Wind Tunnel Studies of Buildings

4.1 Introduction

There are many situations where analytical methods cannot be used to estimate certain types of wind loads and their associated structural responses. For example, the aerodynamic shape of the building is uncommon or the building is very flexible so that its motion affects the aerodynamic forces acting on it. In such situations, more accurate estimates of wind effects on the buildings are obtained through model tests in a boundary-layer wind tunnel, where the boundary layer is simulated by means of the 'roughness, barrier and mixing device' recommended by Cook [4.1]. A typical arrangement of the hardware in a wind tunnel is shown in Fig. 4.1.

The simulated wind must match the vertical variation of the mean wind speed and the turbulent intensity of the atmospheric wind (see Chapter 3). The geometric scale chosen to model the building should closely match the length scale used for the depth of the boundary layer and the integral scale of the longitudinal component of the turbulence. Consistent scaling is important to ensure similarity of the spatial and temporal variations of the fluctuating wind-induced pressures. For sharp-edged bodies, it is not necessary to match the Reynolds numbers, however attaining a minimum value for the Reynolds number is an important consideration in selecting the geometric scale. Another consideration in selecting the geometric scale is with regard to blockage. It is important to avoid excessive (more than 10%) blockage of the test section by the model of the building and the surrounding structures. A geometric scale of 1:300 to 1:500 is commonly used in wind load measurements.

In addition to modelling the approaching flow, it is important to include the influence of the immediate surroundings. This is normally achieved by constructing a proximity model which reproduces in block outline form all major buildings within 0.5 km.

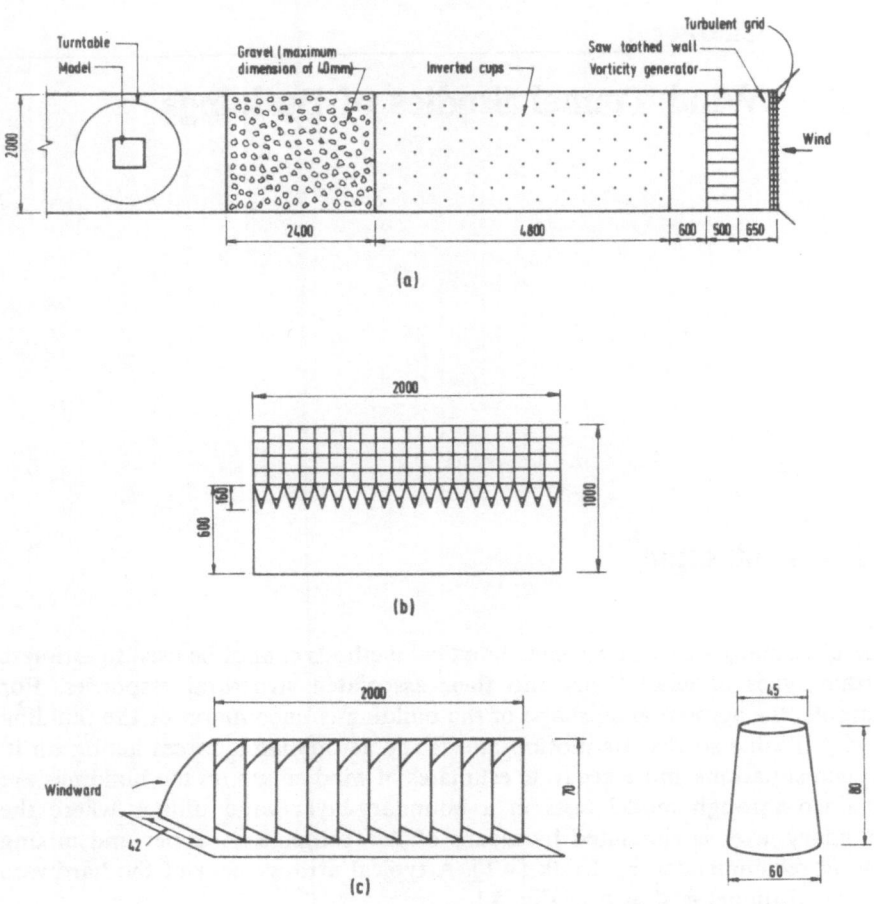

Fig. 4.1. Typical arrangement of hardware for simulation of atmospheric boundary layer: (a) plan view, (b) saw toothed wall tied to the turbulent grid, (c) vorticity generator, (d) inverted cup.

The wind tunnel tests are carried out to determine:

(a) the wind pressures on the exterior surfaces of the building, for cladding design

(b) the overturning moments and shear forces acting on the building, for structural design

(c) the acceleration levels in the building, to ensure human comfort

(d) the changes in the wind environment at ground level, to ensure the safety of pedestrians.

According to the objectives of the test, different types of models are used in wind tunnel tests.

4.2 Rigid Model Studies

Rigid models are used to determine the fluctuating local pressures on the exterior surfaces of the building. It is common to use perspex as the construction material. The exterior features of the building that are considered to be important with regard to the wind flow are simulated to the correct length scale using architectural drawings. Usually a scale of 1:300 to 1:500 is used.

The model is instrumented with a large number of pressure taps (500 to 800) around the model surface to obtain a good distribution of pressures. More tappings are required in regions of high-pressure gradients, such as corners. The pressure tappings are connected by plastic tubing to miniature electronic pressure transducers which can measure the fluctuating pressures. The length of plastic tubing is kept as short as possible to minimize the damping of fluctuating pressures in the tubing. As it is uneconomical to use a single transducer for each pressure tapping, the transducer is mounted onto a pressure-scanning device, such as a scanivalve, which automatically switches the pressure transducer between about 40 to 50 pressure taps, one at a time. Pressure data is acquired by an on-line computer system capable of sampling data at a high speed. Usually a rate of 300 samples/s is used. As the transducer measures only the pressure differentials, the static pressure upstream of the tunnel is used as the reference pressure.

The model is mounted on a turntable in the boundary-layer wind tunnel with the surrounding buildings within a radius of about 500 m. Because of the ease of construction, nearfield features are simulated by polystyrene foam. The turntable provides the facility to test the model at different wind directions by simply rotating the turntable to the desired angle. The pressure measurements are taken for wind directions spaced 10° to 20° apart. For each wind direction, the data are collected for a duration equivalent to 1 hour in the prototype, to obtain stationary values for mean and root mean square pressures. The data record is divided into segments corresponding to 5 to 10 s duration in full scale, and the maximum and minimum values of pressure are calculated for each segment. These individual maximum and minimum values are used in an extreme-value analysis to determine the most probable maximum and minimum values applicable for the whole sample period. The maximum and minimum pressures are expressed as pressure coefficients using the dynamic pressure at the free stream. Knowing the statistical data of a windstorm at the building site, the peak pressures and suctions are computed for the prototype for return periods of 50 years and 100 years. The calculated data are presented in the form of pressure contours or isobars as shown in Fig. 4.2.

In evaluating the peak wind loads on the exterior surface of the prototype, the effects of internal pressure arising from air leakage through openings should be considered. It is also necessary to consider the possibility of window breakage caused by flying debris during a windstorm. As a guide, the resulting internal pressures can be taken to be $\pm 25\,\mathrm{kg/m^2}$ at the base of the building to $\pm 100\,\mathrm{kg/m^2}$ at the roof for a 50-storey building.

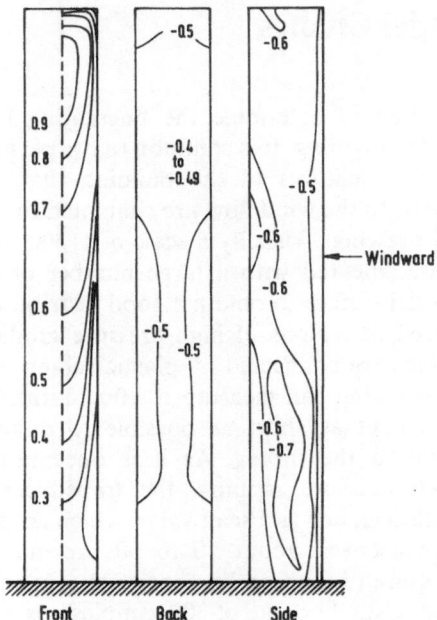

Fig. 4.2. Typical local pressure distribution in terms of pressure coefficients.

For glass design, 1 minute wind loading is commonly used. Since the estimated peak loading is for a duration of 5 to 10 s, a reduction factor has to be used on the peak loads obtained from the wind tunnel tests. Glass manufacturers' recommendations suggest reduction factors of 0.80, 0.94 and 0.97 for annealed float glass, heat strengthened glass and tempered glass, respectively.

For buildings which are not dynamically sensitive to wind action, the rigid model test results can be used to determine the loads for structural design. Generally for buildings with height to width ratios of less than 5, the mean load is obtained from the rigid model tests by integrating the mean pressures over the surface. Then an appropriate gust factor is used to account for the effects of turbulence. The gust factor may be determined from a building code [4.2, 4.3]. The gust factor depends on the averaging period of the mean wind load, terrain roughness in relation to building height, natural frequency of the building, intensity of turbulence and damping of the building.

The procedure to obtain the structural loads is as follows. The mean pressures at various tappings are multiplied by the tributary surface area to obtain the forces acting in two orthogonal directions on elemental areas of the building. These hourly mean forces are multiplied by the appropriate gust factor to include the effects of gusts on the overall loading. By summing the forces acting on elemental areas at particular levels, the distribution of wind loads along the height of the building is obtained. From statics, shear and bending moments at various levels are obtained. The torsional moment can be obtained by appropriate summation of moments about the vertical axis caused by the forces acting on the tributary area of the pressure tappings. As a building in a built-up area may experience reduced mean loads, while the dynamic loads may be quite high, this procedure may underestimate the peak wind loads in some instances and thus it should be used with caution.

4.3 Aeroelastic Model Studies

For buildings which are sensitive to the dynamic action of winds, such as tall, slender and flexible buildings, the body motion may influence the aerodynamic forces acting on the building and hence the resultant response. In such cases, aeroelastic model studies are required to determine the overall mean and dynamic loads, displacements, rotations and accelerations. Aeroelastic model studies may be required under the following situations:

1. when the height to width ratio exceeds 5
2. when the structure is light with a density of the order of 1.5 kN/m³
3. when the fundamental period is long – of the order of 5 to 10 s
4. when the natural frequency of the building in a cross-wind direction is in the neighbourhood of the shedding frequency
5. when the building is torsionally flexible
6. when the building is expected to execute strongly coupled lateral–torsional motion.

For aeroelastic tests, in addition to modelling the properties of atmospheric wind and aerodynamically significant features of the exterior geometry, it is necessary to simulate the mass, stiffness and damping properties of the building.

Thus, equality of the following ratios in model and in full scale needs to be maintained:

$$\frac{\rho_b}{\rho} = \frac{\text{inertia force of building}}{\text{inertia force of flow}} \tag{4.1}$$

$$\frac{E}{\rho V^2} = \frac{\text{elastic force}}{\text{inertia forces of flow}} \tag{4.2}$$

$$\zeta = \frac{\text{dissipating structural forces}}{\text{inertia forces of flow}} \tag{4.3}$$

where ρ, ρ_b, E, V and ζ are respectively the density of air, density of building elastic modulus, wind speed and damping ratio. Measurements on aeroelastic models are carried out at wind speeds that correspond to common events for serviceability and relatively rare events for strength design. Often a 10 year return period is used for the former to obtain information with regard to human comfort and drift, while a 50 or 100 year return period is used for the latter.

Since only the lower modes of vibration contribute significantly to the wind-induced motion of buildings, wind effects can be studied using simple models to simulate the equivalent dynamic properties. The various types of equivalent aeroelastic models are discussed below.

4.3.1 Aeroelastic Model with Linear Mode (Semi-rigid Models)

When the fundamental sway mode of the building can be approximated by a straight line, a rigid model pivoted at the base, as shown in Fig. 4.3, is used as the equivalent aeroelastic model. The pivotal point is chosen in such a way that the corresponding linear mode provides the best fit for the fundamental mode of the prototype. For instance, in the case of a building with a very stiff podium, the pivotal point may be chosen at the intersection of the tower and the podium.

As the model rotates instead of translating, the mass moment of inertia about the pivotal point instead of the mass must satisfy the appropriate model scale. If subscript r denotes the ratio between the prototype and model parameters, that is

$$H_r = \frac{H_p}{H_m}, \qquad I_r = \frac{I_p}{I_m} \tag{4.4}$$

where H is the length, I the mass moment of inertia, and suffixes p and m denote prototype and model respectively, then the appropriate scale for mass

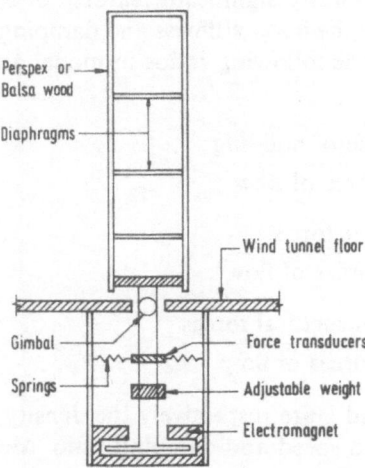

Fig. 4.3. Semi-rigid model with a linear sway mode.

moment of inertia is given by

$$I_r = H_r^5 \tag{4.5}$$

The stiffness scaling is determined from Eq. (4.2) or its equivalent for the lumped parameter system. The equivalent of Eq. (4.2) for vibration in a particular mode with frequency n is

$$\left(\frac{nD}{V}\right)_m = \left(\frac{nD}{V}\right)_p \tag{4.6}$$

where D is the width of the building. From this the velocity scale becomes

$$V_r = \frac{V_p}{V_m} = \left(\frac{n_p}{n_m}\right)\left(\frac{D_p}{D_m}\right) = n_r H_r = \frac{H_r}{T_r} \tag{4.7}$$

where T_r is the time scale and T the period of vibration. When the velocity scale is chosen, the stiffness scale becomes

$$k_r = \frac{I_r}{T_r^2} = H_r^3 H_v^2 \tag{4.8}$$

when k is the rotational stiffness about the pivotal point.

The model mass moment of inertia is determined from the generalized mass of the prototype building for the linear model. Perspex is used to simulate the exterior geometry of the building. To achieve the correct scaling of the mass moment of inertia, the thickness of the outer shell may be varied or mass may be added at the appropriate locations. The damping ratio of the model is adjusted to be the same as the damping ratio for the prototype structure. Typical values for damping are 1% for a steel building and 2% for a concrete building. To construct the model, a convenient model frequency is chosen and from the known prototype frequency the time scale is established. If, for example, $n_m = 24$ Hz, $n_p = 0.18$ Hz and $H_r = 400$, then $V_r = 3$. Thus the design hourly wind speed is reduced by one-third in the wind tunnel. The one hour of full scale event would be compressed to 27 s, since the chosen time scale is 133. Thus the wind tunnel data are collected for 27 s to obtain a stationary sample. From the model test, moment, M, and accelerations, a, are obtained. These are related to the prototype by the following non-dimensional expressions:

$$M_p = M_m\left(\frac{n_p^2 H_p^5}{n_m^2 H_m^5}\right) \tag{4.9}$$

$$a_p = a_m\left(\frac{n_p^2 H_p}{n_m^2 H_m}\right) \tag{4.10}$$

Other quantities of interest are the pressures, P, and shear forces, F, which are given by

$$P_{p} = P_{m}\left(\frac{n_{p}^{2}H_{p}^{2}}{n_{m}^{2}H_{m}^{2}}\right) \tag{4.11}$$

$$F_{p} = F_{m}\left(\frac{n_{p}^{2}H_{p}^{4}}{n_{m}^{2}H_{m}^{4}}\right) \tag{4.12}$$

For a model simulating sway motions in two orthogonal directions and torsional motion about a vertical axis, in addition to the modelling parameters discussed above, the following parameters must be equal in both the model and the prototype:

(a) the critical damping in the longitudinal, lateral and torsional directions
(b) the ratio between the torsional frequency and the longitudinal frequency
(c) the ratio between the lateral frequency and the longitudinal frequency.

A semi-rigid model with three degrees of freedom is shown in Fig. 4.4. Balendra and Nathan [4.4] used five torsion bars to provide stiffness in three orthogonal directions. In the test rig, a single torsion bar is used for the measurement of torsional response, and it is inserted within the model along its vertical axis. The top end of this torsion bar is attached to the model while the botom end is fastened to a perspex disc. The perspex disc is coaxially mounted on an aluminium disc and is free to rotate on top of this disc to vary the angle of attack. The aluminium disc has a vernier scale which allows

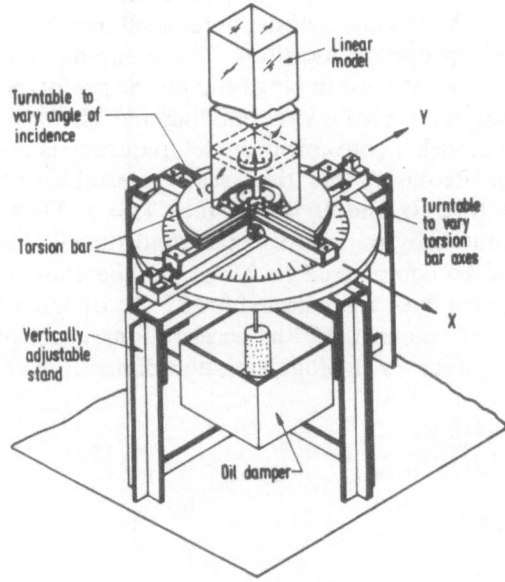

Fig. 4.4. Semi-rigid model with three degrees of freedom.

the angle of attack to be set to an accuracy of one-tenth of a degree. One pair of torsion bars, aligned normal to the along-wind direction for measuring the along-wind response, supports the aluminium disc at one end while the other end is bolted to a cross-beam. Another pair of torsion bars, aligned in the direction of along wind for measuring the across-wind response, supports the cross-beam at one end, while the other end is bolted to a fixed support. The cross-beam is also supported on a pair of ball-bearings to prevent the bending of torsion bars in the along-wind direction. The bending of the torsion bar in the vertical direction is prevented by inserting a ball-bearing between the bottom flange of the torsion bar and the model. The choice of torsional rigidity of the torsion bars determines the natural frequency of the building in any of the three orthogonal directions, and the model can be constrained to oscillate in any particular direction.

A cylindrical perforated damper, mounted onto a second aluminium disc, allows the damping to be simulated in the rocking and torsional modes of oscillations. The damping ratio is changed either by varying the depth of oil in the container or the size of the cylinder. Damping can also be varied by changing the viscosity of the oil. To measure the oscillations of the model, a pair of dynamic strain gauges is attached to each of the torsion bars.

Although the above set-up simulates a torsional mode which has a unit value along the height instead of a linear variation, it nevertheless provides a means of studying the importance of torsional effects and the sensitivity of the dynamic response to eccentricities between the mass centre and the stiffness centre.

4.3.2 Aeroelastic Model with Shear–Flexure Mode

The linear mode used in the semi-rigid model is an approximation to the actual mode shape of typical tall buildings which are constructed with shear walls and frames. When the shapes of a building are complex, with setbacks or major variations in stiffness along the height, the assumption of a linear mode of vibration is not valid. Under lateral loads, the walls deflect in a flexural mode while the frames deflect in a shear mode. As the walls and frames are tied together by the floor slabs, the building deflects in a shear–flexure mode. When the building is modelled as a shear–flexure beam, the governing equation of motion for along-wind displacement of the building $u(z, t)$ is given by

$$m(z)\ddot{u}(z, t) + c(z)\dot{u}(z, t) + EI(z)u''''(z, t) - GA(z)u''(z, t) = f(z, t) \qquad (4.13)$$

where m and c are the mass and damping coefficient per unit height, and EI and GA are respectively the flexural and shear rigidities. Furthermore, $f(z, t)$ is the fluctuating wind load per unit height. In Eq. (4.13), the dots denote the derivative with respect to time t and the primes denote the derivative with respect to z.

Introducing the following non-dimensional parameters

$$u^* = \frac{u}{H}, \qquad z^* = \frac{z}{H} \quad \text{and} \quad t^* = \frac{t}{T} \qquad (4.14)$$

where H is the height of the building and T its period, Eq. (4.13) becomes

$$\ddot{u}^* + \frac{c}{m} T\dot{u}^* + \left(\frac{EI}{m}\right)\frac{T^2}{H^4}(u^*)'''' - \left(\frac{GA}{m}\right)\frac{T^2}{H^2}(u^*)'' = \frac{T^2}{H}\frac{f(z,t)}{m} \qquad (4.15)$$

If subscript r denotes the ratio between the prototype and the model parameters, for Eq. (4.15) to be valid for both prototype and model

$$\frac{T_r^2}{H_r}\frac{f_r}{m_r} = 1, \qquad \frac{c_r T_r}{m_r} = 1 \qquad (4.16a, b)$$

$$\frac{E_r I_r}{m_r}\frac{T_r^2}{H_r^4} = 1, \qquad \frac{G_r A_r}{m_r}\frac{T_r^2}{H_r^2} = 1 \qquad (4.17a, b)$$

Denoting the velocity of the wind as V, then for air density ratio $\rho_r = 1$:

$$f_r = V_r^2 H_r \qquad (4.18a)$$

Since $V_r = H_r/T_r$, substituting Eq. (4.18a) into Eq. (4.16a) leads to

$$m_r = H_r^2 \qquad (4.18b)$$

Since $c_r/m_r = \zeta_r \omega_r$, where ζ_r is the damping ratio and ω_r the frequency, Eq. (4.16b) yields

$$\zeta_r = 1 \qquad (4.18c)$$

Substituting Eq. (4.18b) into Eqs. (4.17a) and (4.17b) yields

$$E_r I_r = V_r^2 H_r^4, \qquad G_r A_r = V_r^2 H_r^2 \qquad (4.19a, b)$$

If the aeroelastic model is constructed to satisfy the mass and stiffness distribution according to Eqs. (4.18b), (4.19a) and (4.19b), then it would have a mode shape similar to the prototype, and the ratio between its frequency and that of the prototype would be the same as the chosen time scale. For ease of construction, the frame elements may be neglected provided the sizes of the flexural elements are adjusted to reflect the shear mode [4.5]. To accomplish this, the prototype should be re-analysed with flexural elements and without the shear elements for the same mass. The sizes of the flexural elements may be adjusted by introducing openings and changing the widths so that the prototype has the same frequency and mode-shape as the one with both flexural and shear elements. Two steel bars have been used to simulate the stiffness of the building in the along-wind direction. The stiffness in the across-wind direction is similarly simulated with another two steel bars. By adjusting the positions of these four steel bars in plan, it should be possible to obtain the required torsional stiffness. The mass and exterior geometry of the building are simulated using a non-structural shell made of perspex. For the model to deflect according to the deformation of the steel bars, the outer shell is made of several segments with small gaps between them. Each segment is connected to the steel bars through horizontal diaphragms. The viscoelastic tape used to cover the gaps between the segments of the outer shell provide the damping.

The base overturning moment is measured using strain gauges at the base of the steel bars. The distribution of wind-induced forces along the height can be obtained by measuring the cross power spectral densities of the pressures acting on the model. The modal force is then given by [4.6]

$$S_{F^*}(n) = \int_0^B \int_0^B \int_0^H \int_0^H \phi(z_1)\phi(z_2)S_p(x_1, z_1, x_2, z_2, n)\, dx_1\, dx_2\, dz_1\, dz_1$$

(4.20)

where B is the breadth, H the height, $\phi(z)$ the mode-shape and S_p the cross power spectral density of pressures between two points (x_1, z_1) and (x_2, z_2), which is obtained by measuring the pressure signals from two tappings at the same time using a sample and hold hardware device. The data are then converted into the frequency domain by the fast Fourier transform technique.

The power spectral density of displacement at any height of the model is then determined from

$$S_d(z, n) = \phi^2(z)|\hat{H}(n)|^2 S_{F^*}(n)$$

(4.21)

where $|\hat{H}(n)|$ is the transfer function of the modal equation corresponding to the mode-shape $\phi(z)$, and n_m is the corresponding frequency of vibration.

The root mean square dynamic displacement is given by

$$\sigma_d(z) = \left[\frac{\phi^2(z)}{(m^*)^2(2\pi n_m)^4} \frac{(\pi n_m)}{4\zeta_m} S_{F^*}(n_m) \right]^{1/2}$$

(4.22)

where m^* is the generalized mass and ζ_m the damping ratio for the chosen mode of vibration. The root mean square acceleration at any height is given by

$$\sigma_a(z) = (2\pi n_m)^2 \sigma_d(z)$$

(4.23)

The root mean square shear and the moment at any height z_0 from the base of the model is obtained from

$$Q(z_0) = (2\pi n_m)^2 \int_{z_0}^H m(z)\sigma_d(z)\, dz$$

(4.24)

$$M(z_0) = (2\pi n_m)^2 \int_{z_0}^H (z - z_0)m(z)\sigma_d(z)\, dz$$

(4.25)

4.3.3 Aeroelastic Model with Coupled Modes

For more complex buildings where torsional modes are important or in situations where the modes of vibrations are strongly coupled due to eccentricity between the mass centre and the elastic centre, a discrete model with several

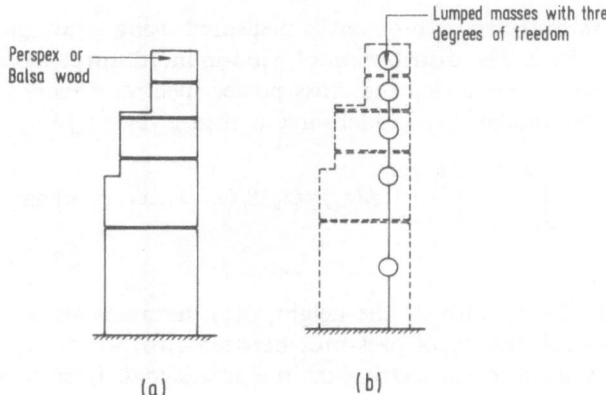

Fig. 4.5. Schematic diagram of discrete aeroelastic model: (a) exterior view of the model, (b) lumped mass representation.

lumped masses interconnected by flexible columns is used [4.7, 4.8]. The building is divided into several zones and the mass of each zone is concentrated at the centre of the zone on a horizontal diaphragm with two translational and one rotational (about the vertical axis) degrees of freedom. The horizontal diaphragms are connected by flexible columns, and then the entire mechanical system is enclosed by a non-structural shell which simulates the exterior geometry of the building (Fig. 4.5). The shell is made discontinuous to allow relative movements between the masses. If the diaphragms are made stiff, then the model simulates the behaviour of a shear building where the axial deformation of the columns and rotation of horizontal girders are neglected. With considerable fabrication effort, Isyumov [4.9] has simulated the girder rotation and axial deformation of the columns (Fig. 4.6).

In studies of most tall buildings, four lumped masses with 12 degrees of freedom are found to be sufficient. Inclusion of higher sway modes in

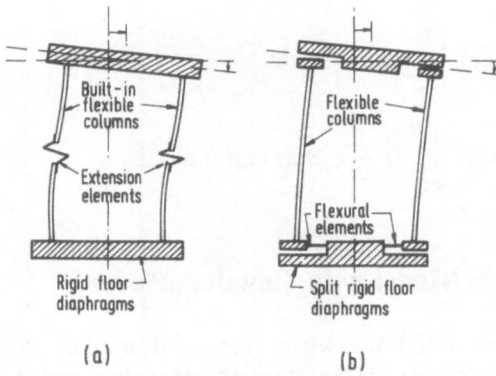

Fig. 4.6. Aeroelastic model with: (a) axial deformation of columns, (b) rotation of floor diaphragms.

multi-degree-of-freedom system simulation improves the results only marginally, as the response is predominantly in the two fundamental sway modes.

4.4 High-frequency Force Balance Model

An aeroelastic model provides comprehensive information on the dynamic loads and motion of the prototype. However, construction of an aeroelastic model is complex and costly, and cannot be carried out until the essential structural features, such as the distribution of stiffness and mass of the prototype are finalized. The high-frequency force balance technique provides an alternative method which is more economical and time efficient. In this method, the generalized wind-induced forces in a building with a linear mode-shape are determined by measuring the dynamic base moment acting in a rigid model simulating the geometry of the building. The model is mounted on a highly sensitive stiff force balance which measures the base overturning moment. The frequencies of the model and the balance are chosen to be sufficiently high, so that there are no distortions in the dynamic wind loads due to resonance in the frequency range of interest. The power spectrum of the measured base moment is the same as the power spectrum of the generalized force corresponding to a linear mode. From the power spectrum of the generalized force, the root mean square dynamic displacement, acceleration, shear and moment are determined analytically using Eqs. (4.21) to (4.25).

The high-frequency force balance technique allows the dynamic response of several alternative structural systems to be evaluated economically. However, it is applicable only when the motion of the building does not affect the aerodynamic forces, as the model does not simulate the dynamic properties of the building.

4.5 Pedestrian Wind Studies

During high winds, when the path of the wind is blocked by the broad side of a tall, flat building, the tendency of the wind is to drift in a vertical direction rather than to go around the building at the same level. As such, some of the wind will be deflected upward, but most will spiral to the ground creating a strong wind at the pedestrian level. Thus walking in the neighbourhood of skyscrapers may become extremely unpleasant and some times dangerous for pedestrians. The presence of other tall buildings nearby may aggravate the situation. Thus for successful design of buildings at the pedestrian level, it is important to estimate the wind conditions in outdoor areas of buildings and building complexes through model studies to ensure safety and comfort. These studies require a geometrically scaled model which includes aerodynamically significant details of the building. Significant structures nearby and any important topographic features also need to be included.

When the model is tested in the wind tunnel under the simulated atmospheric wind, wind speed measurements are made at various locations at the pedestrian level and compared with a set of standard acceptance criteria. If unacceptable pedestrian-level wind speeds are detected, remedial procedures can be suggested through these model studies. The effects of erecting a proposed building within a cluster of existing buildings are obtained by comparing the pedestrian-level wind speeds with and without the proposed building.

References

4.1. Cook N J, Wind tunnel simulation of the adiabatic atmospheric boundary layer by roughness, barrier and mixing device methods, *Journal of Wind Engineering and Industrial Aerodynamics* 1978; 3: 157–176.
4.2. *NBC: National Building Code of Canada*, National Research Council of Canada, Ottawa, Canada, 1985.
4.3. ANSI A58.1: *Minimum Design Loads for Buildings and Other Structures*, American National Standard Institute, New York, 1982.
4.4. Balendra T and Nathan G K, Longitudinal, lateral and torsional oscillations of a square section tower model in an atmospheric boundary layer, *Journal of Engineering Structures* 1987; 4: 217–288.
4.5. Balendra T, Cheong H F and Lee S L, An aeroelastic model for shear wall–frame buildings, *International Journal of Structures* 1991; 11: 131–143.
4.6. Cheong H F, Balendra T, Chew Y T, Lee T S and Lee S L, An experimental technique for distribution of dynamic wind loads on tall buildings, *Journal of Wind Engineering and Industrial Aerodynamics* 1992; 40: 249–261.
4.7. Skilling J B, Tschanz T, Isyumov N, Loh P and Devenport A G, Experimental studies, structural design and full-scale measurements for the Columbia Seafirst Centre, *Proceedings of Building Motion in Wind, ASCE Convention*, Seattle, Washington, 1986, pp 1–22.
4.8. Templin J T and Coper K R, Design and performance of a multi degree of freedom aeroelastic building model, *Journal of Wind Engineering and Industrial Aerodynamics* 1981; 8: 157–175.
4.9. Isyumov N, The aeroelastic modelling of tall buildings, *International Workshop on Wind Tunnel Modelling Criteria and Techniques*, National Bureau of Standards, Gaithersburg, Maryland, 1982, pp 373–407.

Chapter 5

Analysis of the Behaviour of Buildings During Earthquakes

5.1 Earthquake Loading

Earthquakes are vibrations of the earth's surface caused by sudden movements of the earth's crust which consists of a number of thick rock plates that float on the earth's molten mantle. The plates drift on convection currents generated by hot spots deep within the earth, and deform as they move, owing to interlocking at plate boundaries. As a result, stress builds up and when the shear stress exceeds the strength of the rock, a rupture occurs along the fault line in the rock and energy is released in the form of seismic waves. The origin of the fracture is known as the *focus* of the earthquake (Fig. 5.1). Two kinds of body waves are propagated from the focus. The first is the compressional or *P* wave, which is propagated as an expanding sphere of disturbance. The second is the *S* wave which is characterized by shearing distortion without any volumetric change. The point on the surface directly above the focus is called the *epicentre* of the earthquake. When body waves strike the free surface, they give rise to two kinds of surface wave. The first are called Love waves, and consist of a horizontal motion of the surface transverse to the direction of propagation. The second are called Rayleigh waves, in which surface particles

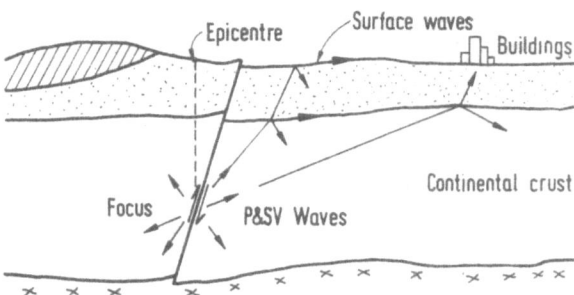

Fig. 5.1. Types of seismic waves.

move in vertical retrograde elliptical orbits. Body-wave amplitudes decay at the rate of r^{-2}, where r is the radial distance from the focus, whereas surface-wave amplitudes decay at the rate of $r^{-1/2}$. As the latter decay much less rapidly, such earthquakes may be felt hundreds of miles from their epicentres, and contribute more to earthquake damage.

Ground motions caused by seismic waves are measured by the strong motion accelerograph, a device sensitive to the ground motions most likely to affect structures. It records three components of ground acceleration, two horizontal and one vertical. An example of a strong motion record is shown in Fig. 5.2. The recorded horizontal acceleration, after correction for instrument characteristics, is digitized and then integrated to obtain the velocity and displacement time histories shown in Fig. 5.2.

Since the foundation is the point of contact between the building and the earth, the seismic waves act on the building by shaking the foundation back and forth. The mass of the building resists this motion, setting up inertia forces throughout the structure. Vertical inertia forces are generally unimportant however, since buildings are already designed for static vertical loading and hence are strong in this direction. Thus only horizontal inertia forces need be considered, and these may exceed the wind forces acting on a structure.

The magnitude of the horizontal inertia force depends on the building's mass, the ground acceleration and the type of structure. If a building and its foundation were rigid, it would have the same acceleration as the ground, and the peak lateral force would be mass times peak acceleration. In reality, this is never the case, since all buildings are flexible to some degree. For a structure that deforms only slightly, thereby absorbing some energy, the force may be

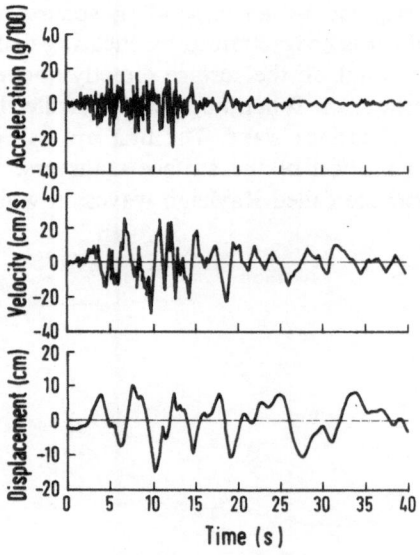

Fig. 5.2. Earthquake strong motion record.

less than the product of mass and acceleration. But a very flexible structure, having a natural period near that of the ground motion, may be subject to a much larger force. Thus the magnitude of the lateral force on a building depends not only on the peak acceleration of the ground motion but also on the frequency content. The importance of the frequency content was well illustrated by the 1985 Mexico City earthquake. The ground motion at Mexico City had a predominant period of 2 s. This coincided with the natural period of vibration of buildings in the medium height range, typically six to twenty storeys, many of which collapsed or suffered serious damage, whereas other buildings were practically unaffected.

5.2 Response Spectrum Analysis of SDOF

The most widely adopted method of studying the frequency content of earthquakes is by the response spectrum technique. The response spectrum of an earthquake depicts the variation of peak dynamic response of a single-degree-of-freedom system (SDOF) for different values of its natural frequency, and for a particular damping ratio. Consider a single-storey shear building subjected to ground motion $v_g(t)$, as shown in Fig. 5.3. The mass is not subjected to any external loading, thus for equilibrium

$$F_I + F_D + F_S = 0 \tag{5.1}$$

where F_I is the inertia force, F_D the damping force and F_S the elastic restoring force. The inertia force depends on the total acceleration of the mass, $\ddot{v}_t$, which is given by

$$\ddot{v}_t(t) = \ddot{v}_g(t) + \ddot{v}(t) \tag{5.2}$$

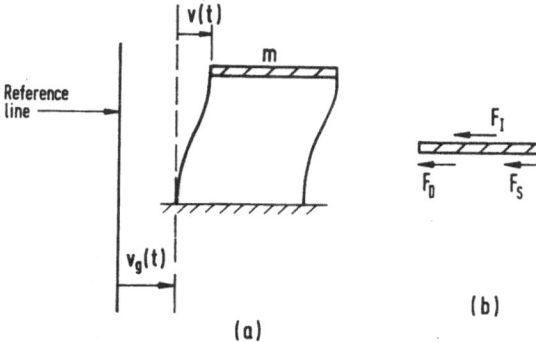

Fig. 5.3. (a) One-storey shear building subjected to ground motion and (b) forces acting on the floor masses.

Thus the governing equation of motion becomes

$$m\ddot{v} + c\dot{v} + kv = P_{\text{eff}}(t) \tag{5.3}$$

where

$$P_{\text{eff}}(t) = -m\ddot{v}_{\text{g}} \tag{5.4}$$

is the effective force resulting from the ground motion. The other symbols, m, c, and k, are as defined in Chapter 1.

The solution to Eq. (5.3) for zero initial condition is given by the Duhamel integral. Neglecting the small difference between the damped frequency and the undamped frequency, and also the negative sign in Eq. (5.4), as it is of little interest in earthquake response analysis, we have

$$v(t) = \frac{1}{\omega}\left[\int_0^t \ddot{v}_{\text{g}}(\tau)\, e^{-\zeta\omega(t-\tau)} \sin \omega(t-\tau)\, d\tau\right] \tag{5.5}$$

The above equation gives the time history response of relative displacement which is a function of the natural frequency, ω, and the damping ratio, ζ, of the system.

The maximum relative displacement is determined from the maximum value of the integral which is denoted as $S_v(\omega, \zeta)$ and is called the *spectral pseudo velocity*:

$$S_v(\omega, \zeta) = \left[\int_0^t \ddot{v}_{\text{g}}(\tau)\, e^{-\zeta\omega(t-\tau)} \sin \omega(t-\tau)\, d\tau\right]_{\text{max}} \tag{5.6}$$

According to Eq. (5.6), the maximum relative displacement is given by

$$[v(t)]_{\text{max}} = \frac{1}{\omega} S_v(\omega, \zeta) = S_d(\omega, \zeta) \tag{5.7}$$

where S_d is the spectral displacement. The maximum absolute acceleration or the spectral acceleration, S_a, may be expressed as

$$S_a(\omega, \zeta) = \omega^2 S_d(\omega, \zeta) \tag{5.8}$$

For a given earthquake record, by assuming a specific value of damping for the SDOF oscillator, it is possible to plot the variation of S_a, S_v and S_d with the natural period or frequency of the oscillator. The graph showing such a variation is called the *response spectrum* of the earthquake motion. For example, the acceleration response spectrum of the El-Centro earthquake is shown in Fig. 5.4. The sharp peaks and valleys in the response spectrum are due to local resonances in the ground motion record. Such irregularities are not of fundamental significance and may be smoothed out by averaging the response spectra of a number of different earthquake records which are normalized to a particular intensity level, to obtain the design spectrum. In Fig. 5.5, a design spectrum normalized to a peak ground acceleration of 1g is presented.

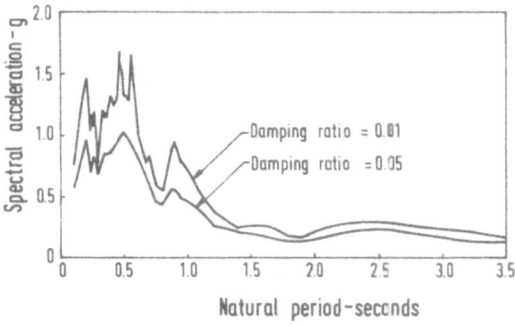

Fig. 5.4. Acceleration response spectrum for El-Centro earthquake of 18 May 1940 (N–S component).

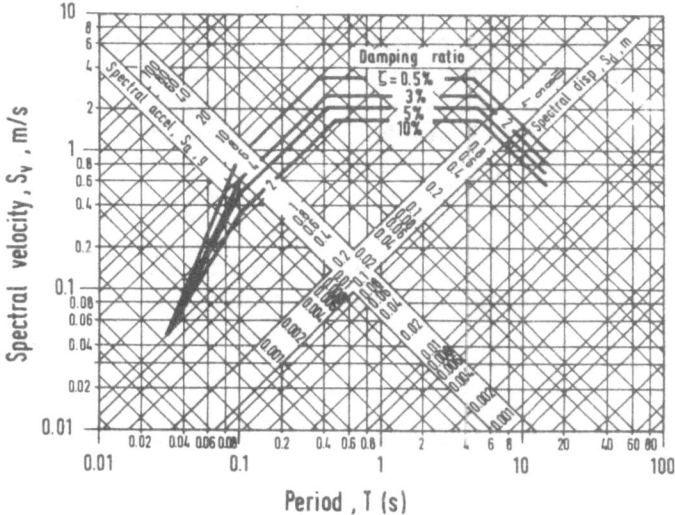

Fig. 5.5. Design spectrum normalized to a peak ground acceleration of $1g$.

5.3 Response Spectrum Analysis of MDOF

When a shear building with N storeys, as shown in Fig. 5.6, is subjected to base excitation, the governing equations of motion take the form

$$[M]\{\ddot{v}\} + [C]\{\dot{v}\} + [K]\{v\} = -[M]\{1\}\ddot{v}_g \tag{5.9}$$

where $\ddot{v}_g$ is the ground acceleration, $\{v\}$ is the displacement vector of floor masses with respect to the base, $\{1\}$ represents a unit vector and $[M]$, $[C]$ and $[K]$ are as defined in Chapter 1 (section 1.3).

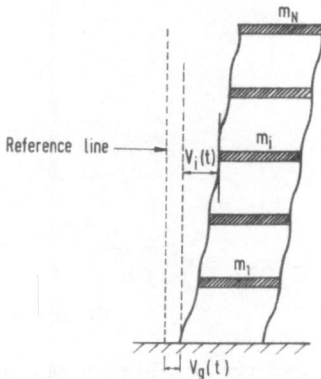

Fig. 5.6. *N-storey shear building subjected to ground motion.*

For a linear system with proportional damping, Eq. (5.9) can be solved effectively by modal analysis. For earthquake response analysis, the modal analysis technique becomes much more efficient, because ground motions tend to excite strongly only the lowest few modes of vibration. Thus, Eq. (5.9) can be reduced to n modal equations of the form

$$\ddot{q}_r + 2\zeta_r\omega_r\dot{q}_r + \omega_r^2 q_r = -\ddot{v}_g\gamma_r, \qquad r = 1, 2, \ldots, n \qquad (5.10)$$

where $n \ (\leq N)$ is the number of significant modes and the modal coordinate q_r is related to the displacement of the i-th mass as follows:

$$v_i = \sum_{r=1}^{n} \Phi_{ir} q_r \qquad (5.11)$$

in which Φ_{ir} is the i-th component of the r-th mode-shape vector. Furthermore, the modal participation factor γ_r in Eq. (5.10) is given by

$$\gamma_r = \frac{\sum_{i=1}^{N} m_i \Phi_{ir}}{\sum_{i=1}^{N} m_i \Phi_{ir}^2} \qquad (5.12)$$

Equation (5.10) represents the equation of motion of a SDOF system and the response is obtained from the Duhamel integral as

$$q_r(t) = -\frac{\gamma_r}{\omega_r} \int_0^t \ddot{v}_g(\tau) e^{-\zeta_r\omega_r(t-\tau)} \sin \omega_r(t - \tau) \, d\tau \qquad (5.13)$$

The time history response of the i-th mass is then determined from Eq. (5.11) as

$$v_i(t) = \underbrace{\Phi_{i1}q_1(t)}_{\text{1st modal response}} + \underbrace{\Phi_{i2}q_2(t)}_{\text{2nd modal response}} + \cdots \qquad (5.14)$$

To determine the maximum response, according to Eq. (5.6), the maximum response of the modal coordinate is first obtained from Eq. (5.3) as

$$(q_r)_{max} = \gamma_r \frac{S_v}{\omega_r}(\omega_r, \zeta_r)$$

$$= \gamma_r S_d(\omega_r, \zeta_r) \tag{5.15}$$

where S_d is the spectral displacement corresponding to modal frequency ω_r and modal damping ratio ζ_r in the r-th mode. According to Eq. (5.14), the maximum relative displacement of the i-th mass in the r-th mode is given by

$$(v_{ir})_{max} = \Phi_{ir}(q_r)_{max} \tag{5.16}$$

The maximum interstorey drift between any two storeys in the r-th mode may be obtained from

$$(v_{ir} + v_{i-1,r})_{max} = (\Phi_{ir} - \Phi_{i-1,r})\gamma_r S_d(\omega_r, \zeta_r) \tag{5.17}$$

The maximum inertia force in the r-th mode at the i-th level is given by

$$(F_{ir})_{max} = m_i \omega_r^2 (v_{ir})_{max}$$

$$= m_i \Phi_{ir} \gamma_r S_a(\omega_r, \zeta_r) \tag{5.18}$$

and the maximum dynamic shear at the k-th storey and base shear in the r-th mode are obtained by summing the inertia forces induced:

$$(V_{kr})_{max} = \gamma_r S_a(\omega_r, \zeta_r) \sum_{i=k}^{N} m_i \Phi_{ir} \tag{5.19a}$$

$$(Q_r)_{max} = \gamma_r S_a(\omega_r, \zeta_r) \sum_{i=1}^{N} m_i \Phi_{ir} \tag{5.19b}$$

where S_a is the spectral acceleration corresponding to modal frequency ω_r and damping ratio ζ_r.

According to Eq. (5.14), the total response is the sum of the modal responses. Since the maximum response in each mode would occur at different times, it would be conservative to superpose the maximum responses obtained for each mode of vibration. The recommended procedure to determine the most probable maximum response is by the square root sum square (SRSS) method [5.1]. Accordingly, the maximum response R is given by

$$R = \left[\sum_{r=1}^{n} (R_r)_{max}^2 \right]^{1/2} \tag{5.20}$$

where $(R_r)_{max}$ is the maximum value of the quantity R (displacement, drift, base shear, etc.) in the r-th mode.

The SRSS method may lead to significant errors if the modal frequencies are closely spaced. For such cases, the CQC (complete quadratic combination) method is recommended [5.2]. Accordingly

$$R = \left[\sum_{r=1}^{n} \sum_{s=1}^{n} R_r \beta_{rs} R_s \right]^{1/2} \tag{5.21}$$

where R_r and R_s are the maximum responses in the r-th and s-th modes, respectively, and the cross modal coefficient β_{rs} obtained from random vibration theory is given by

$$\beta_{rs} = \frac{8(\zeta_r\zeta_s)^{1/2}(\zeta_r + \rho\zeta_s)\rho^{3/2}}{(1 - \rho^2)^2 + 4\zeta_r\zeta_s\rho(1 + \rho^2) + 4(\zeta_r^2 + \zeta_s^2)\rho^2} \tag{5.22}$$

where $\rho = \omega_s/\omega_r$, and ζ_r and ζ_s are the damping ratios in the r-th and s-th modes.

The elastic response spectrum technique described above would give the maximum response if the structure remained elastic. However, since buildings are designed with a certain amount of ductility to absorb energy through inelastic deformation in the event of a severe earthquake, the recommended procedure by the Canadian Code of Practice [5.3] is to scale down the response obtained from the elastic response spectrum by a factor equal to the ratio between the equivalent lateral base shear given in the code (which accounts for ductility, see section 5.5) and the dynamic base shear obtained from the elastic response spectrum technique.

Example 5.1. Compute the maximum floor displacement and storey shear for a six-storey shear building using the elastic response spectrum given in Fig. 5.5 for 5% damping. The peak ground acceleration is to be taken as 0.05g. The mass at various floors, the first three modes and the corresponding periods of vibrations are given in Table 5.1.

Solution. The maximum floor displacement and the maximum inertia force at floor level i in the r-th mode of vibration are given by

$$(v_{ir})_{\max} = \Phi_{ir}\frac{\gamma_r}{\omega_r^2}S_a(\omega_r, \zeta_r)$$

$$(F_{ir})_{\max} = m_i\Phi_{ir}\gamma_rS_a(\omega_r, \zeta_r)$$

The modal participation factor computed from

$$\gamma_r = \frac{\sum_{i=1}^{N} m_i\Phi_{ir}}{\sum_{i=1}^{N} m_i\Phi_{ir}^2}$$

Table 5.1. Properties of building for Example 5.1.

Floor level i	Mass (kg) m_i	Mode 1 Φ_{i1}	Mode 2 Φ_{i2}	Mode 3 Φ_{i3}
6	1.2×10^6	0.550	−0.520	0.455
5	1.2×10^6	0.520	−0.252	−0.135
4	1.2×10^6	0.456	0.140	−0.560
3	1.2×10^6	0.365	0.460	−0.254
2	1.2×10^6	0.254	0.560	0.372
1	1.2×10^6	0.120	0.368	0.520
Period (s) T_r		0.60	0.20	0.10

for the first three modes are

$$\gamma_1 = 2.28$$

$$\gamma_2 = 0.746$$

$$\gamma_3 = 0.393$$

The spectral acceleration for the first three modes are determined from Fig. 5.5 as

$$S_a(T_1, \zeta_1) = S_a(0.60, 0.05) = 2.13g$$

$$S_a(T_2, \zeta_2) = S_a(0.20, 0.05) = 3g$$

$$S_a(T_3, \zeta_3) = S_a(0.10, 0.05) = 3g$$

Since the spectrum in Fig. 5.5 is for a peak ground acceleration of $1.0g$, the above spectral values must be multiplied by 0.05 to obtain the spectral values corresponding to a peak ground acceleration of $0.05g$. Thus, for mode 1:

$$(v_{i1})_{max} = \Phi_{i1}(2.28)\left(\frac{0.60}{2\pi}\right)^2 (2.13g)(0.05)(1000)$$

$$= 21.7\Phi_{i1} \text{ mm}$$

$$(F_{i1})_{max} = m_i \Phi_{i1}(2.28)(2.13g)(0.05)$$

$$= 2.38 m_i \Phi_{i1} \text{ N}$$

For mode 2:

$$(v_{i2})_{max} = \Phi_{i2}(0.746)\left(\frac{0.20}{2\pi}\right)^2 (3g)(0.05)(1000)$$

$$= 1.1\Phi_{i2} \text{ mm}$$

$$(F_{i2})_{max} = m_i \Phi_{i2}(0.746)(3g)(0.05)$$

$$= 1.1 m_i \Phi_{i2} \text{ N}$$

For mode 3:

$$(v_{i3})_{max} = \Phi_{i3}(0.393)\left(\frac{0.10}{2\pi}\right)^2 (3g)(0.05)(1000)$$

$$= 0.146\Phi_{i3} \text{ mm}$$

$$(F_{i3})_{max} = m_i \Phi_{i3}(0.393)(3g)(0.05)$$

$$= 0.578 m_i \Phi_{i3} \text{ N}$$

The maximum modal displacements and inertia forces at various floor levels are tabulated in Table 5.2 for the first three modes. From the inertia forces, the maximum modal storey shears V_{ir} are computed from statics. Since the

Table 5.2. Computation of floor displacement and storey shear for Example 5.1.

Floor	Storey	$(v_1)_{max}$ (mm)	$(v_2)_{max}$ (mm)	$(v_3)_{max}$ (mm)	v_{max} (mm)	$(F_{i1})_{max}$ ($\times 10^6$ N)	$(V_{i1})_{max}$ ($\times 10^6$ N)	$(F_{i2})_{max}$ ($\times 10^6$ N)	$(V_{i2})_{max}$ ($\times 10^6$ N)	$(F_{i3})_{max}$ ($\times 10^6$ N)	$(V_{i3})_{max}$ ($\times 10^6$ N)	V_{max} ($\times 10^6$ N)
6		11.64	−0.572	0.066	11.65	1.57	1.57	−0.686	−0.686	0.316	0.316	1.74
5	6	11.28	−0.277	−0.02	11.28	1.49	3.06	−0.333	−1.02	−0.094	0.222	3.23
4	5	9.90	0.154	−0.082	9.90	1.3	4.36	0.185	−0.834	−0.389	−0.167	4.44
3	4	7.92	0.506	−0.037	7.94	1.04	5.40	0.607	−0.227	−0.176	−0.343	5.42
2	3	5.51	0.616	0.054	5.54	0.73	6.13	0.739	0.512	0.258	−0.085	6.15
1	2	2.60	0.405	0.076	2.63	0.34	6.47	0.486	0.998	0.361	0.276	6.55

frequencies are well separated, the modal maxima are combined using the SRSS method to obtain the most probable maximum floor displacements and storey shears.

5.4 Site and Soil–Structure Interaction Effects

When seismic waves propagate through the soil layer overlying rock, the soil layer filters the high-frequency components of rock motion and introduces a large proportion of longer-period components centred around the site period. Thus the surface motion could be significantly different from the rock motion both in magnitude and frequency content owing to the site effect. Seed *et al.* [5.4] have obtained earthquake response spectra for different soil conditions (Fig. 5.7). It can be seen that the effect of the subsoil is to magnify the ground motion around the site period. Thus if the building period happened to be in the neighbourhood of the site period, there will be a resonance effect at the soil–structure period. Balendra and Heidebrecht [5.5] proposed a foundation factor that could be used in building codes to account for the amplification of base shear in buildings constructed on stiff soil, deep cohesionless soil and soft soil, considering both the site and soil–structure interaction effects. The latter introduces additional degrees of freedom to the footing (translation and rocking motion for two-dimensional structures) to account for deformation of soil around the footing during energy transmission between the soil and the structure. The translation component is important for low buildings whereas the rocking component is more significant for tall buildings. However, soil–structure interaction has another important effect – energy dissipation due to radiation of waves into the soil medium. This increases the damping of the system substantially. Consequently, the effect of soil–structure interaction on the base shear can be conservatively neglected [5.6]. Because of rocking, the

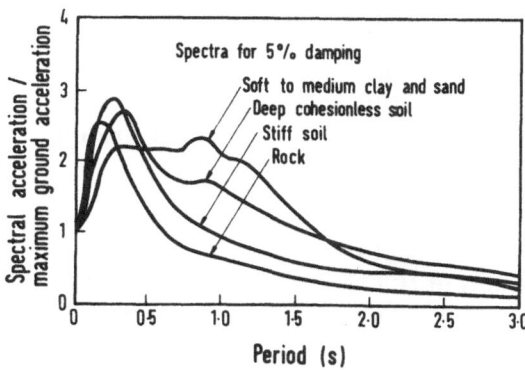

Fig. 5.7. Average acceleration spectra for different site conditions.

deflection increases for taller buildings, but the effect is small. Thus a building supported on a soil layer could be analysed by the method described in the preceding section, using the appropriate response spectrum which accounts for the site effect. In the absence of such a spectrum, Balendra *et al.* [5.7] modified the power spectral density of the rock motion to account for the site effect and obtained the power spectral density of the ground motion. From this, the most probable maximum response of buildings was obtained, using random vibration theory.

5.5 Equivalent Lateral Load Analysis

In this method, the dynamics problem is reduced to a statics problem. The base shear, Q, induced during earthquake motion is expressed as a fraction of the weight of the building, namely

$$Q = C_s W \tag{5.23}$$

where W is the total dead load plus part of the live load that could possibly be present, and C_s is the seismic coefficient, which is expressed in several different ways in the various building codes. However, a useful general version to illustrate the salient features is given by Smith [5.8] as

$$C_s = ASI/R \tag{5.24}$$

where A is the site-dependent effective peak acceleration (as a fraction of g), S the seismic response factor, I the importance factor and R a factor related to the ductility of the structure.

The effective peak acceleration, A, is related to the seismicity of the site. The appropriate value of A may be obtained from a seismic zoning map given in building codes. Since the spectral velocity is almost constant over the practical range of building periods, the value of A is related to the peak velocities given in seismic zone maps. The seismic response factor, S, takes into account the influence of the period of the building and the effect of soil condition. The approved variation of S (Fig. 5.8) with period is given by the Applied Technology Council [5.6]. The three types of soil condition, S_1, S_2, and S_3, are described as follows:

S_1: Rocks of any type with shear wave velocities greater than about 750 m/s; shallow stable deposits of sands, gravels or stiff clays overlying rock.
S_2: Deep cohesionless or stiff clay deposits of 50 to 100 m depth.
S_3: Soft to medium stiff clays and sands with depths of 10 m or more.

The period of the building, T (in seconds), required to estimate the seismic response factor may be calculated as follows [5.8]:

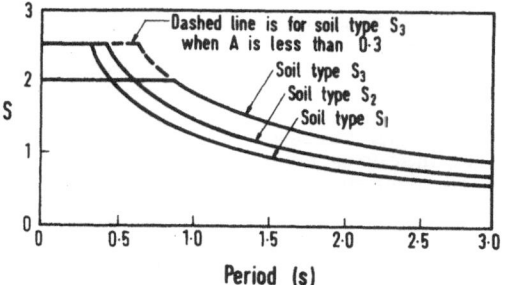

Fig. 5.8. Seismic response factor for $A = 1.0$.

1. For moment-resisting frames of N storeys

$$T = 0.1N \qquad (5.25)$$

2. For shear-wall buildings and braced steel frames of height H (m) and breadth B (m) in the direction of the seismic forces

$$T = 0.09H/B^{1/2} \qquad (5.26)$$

The importance factor I is introduced to ensure certain categories of buildings are designed for greater levels of safety;

$I = 1.5$ is used for essential buildings, such as hospitals, fire stations, etc.
$I = 1.2$ is used for places for assembly, such as schools, theatres, etc.
$I = 1.0$ is used for other types of building.

The factor R accounts for the ductility capacity of the structure, which is defined as the ratio of the displacement at ultimate load to the displacement at yield, as illustrated in Fig. 5.9. It also reflects the ability of different structural forms to develop alternative load paths. The values of R for different structural systems are indicated in Table 5.3.

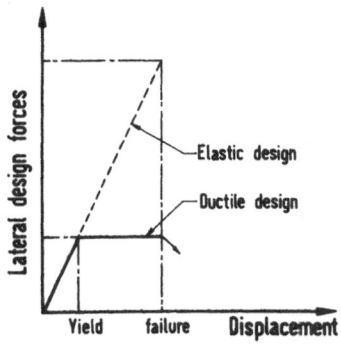

Fig. 5.9. Lateral forces in ductile design and elastic design.

Table 5.3. R-factors used to account for ductility.

Type of structural system	Seismic-resisting system	R
Building frame system An essentially complete space frame providing support for vertical loads, where seismic resistance is provided by shear walls or braced frames	Reinforced concrete shear walls	5.5
	Braced frames	5.0
	Unreinforced masonry shear walls	1.5
Moment-resisting frame Similar to above except that seismic force resistance is provided by special moment-resisting frames capable of resisting the total seismic forces	Reinforced concrete frames	7.0
	Steel frames	8.0
Dual system Seismic resistance is provided by a combination of special moment-resisting frames (at least 25% of seismic forces) and shear walls or braced frames	Reinforced concrete shear walls	8.0
	Braced frames	6.0
Inverted pendulum structure The framing acts essentially as a vertical cantilever resisting seismic and vertical loads, e.g. elevated storage tanks	Special moment-resisting frame of steel or reinforced concrete	2.5

The base shear calculated from Eq. (5.23) is distributed along the height of the building. The distribution of lateral forces along the height of the building depends on the mode-shape of the building. Thus the base shear obtained from Eq. (5.23) may be distributed as

$$F_i = Q\left(\frac{m_i h_i}{\sum_i m_i h_i}\right) \tag{5.27}$$

where F_i is the lateral force at level i, m_i the mass at level i and h_i the height of level i from the base. The above distribution assumes the fundamental mode to be linear. This assumption is valid for regular short-period buildings as the influence of higher modes is small, and the mode-shape is approximately linear. However, for long-period buildings, the influence of higher modes can be significant, and in regular tall buildings the fundamental mode lies approximately between a straight line and a parabola. In order to reflect the contribution of higher modes, the Applied Technology Council [5.6] has proposed the following distribution:

$$F_i = Q\frac{m_i h_i^k}{\sum_i m_i h_i^k} \tag{5.28}$$

where the value of k is taken as

$k = 1$ for $T < 0.5$ s

$k = 2$ for $T > 2.5$ s

For periods between 0.5 and 2.5 s, k is obtained by linear interpolation.

As the ground motion could act in any direction with respect to the principal orthogonal axes, the resultant response is obtained by combining the response due to 100% of the lateral forces along one axis with the response due to 30% of the lateral forces along the orthogonal axis.

The equivalent lateral force analysis is not applicable to buildings of irregular form, irregular mass or stiffness distribution, or of unusual shape. Furthermore, the method is not useful when information on the motion of the building is required. For such cases, dynamic analysis is recommended.

5.6 Inelastic Response Analysis

Inelastic response analysis is important in the earthquake design of buildings because a structure is permitted to go into the inelastic range if it is subjected to strong but rare earthquake motion. Thus in order to determine the response of buildings to strong earthquakes, a proper inelastic analysis needs to be carried out. To do this, an appropriate analytical model defining the restoring force-displacement characteristics of the structure is required. For steel frames, either the bilinear hysteretic model in Fig. 5.10a or the elasto plastic model (slope of AB in Fig. 5.10a is zero) is used.

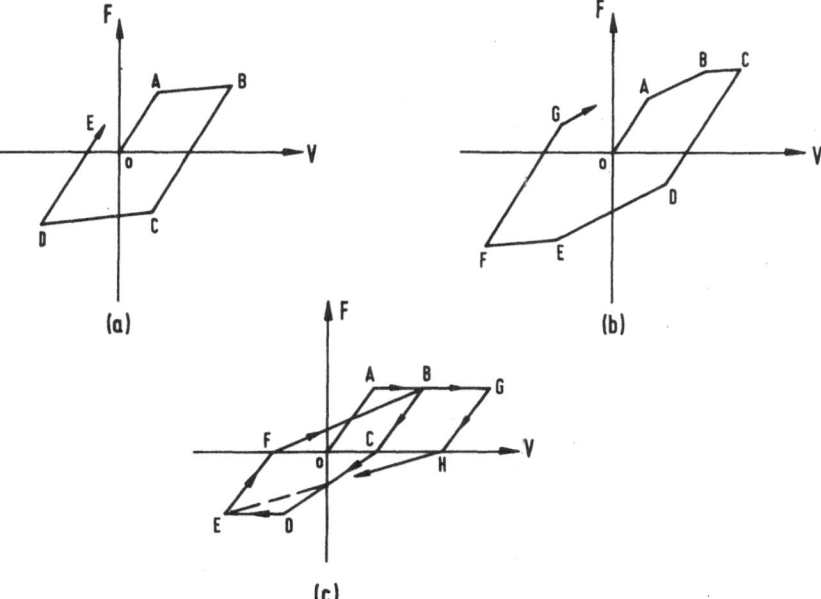

Fig. 5.10. Force–displacement hysteresis: (a) bilinear model, (b) trilinear model, (c) model with stiffness degradation.

A trilinear model applicable to reinforced concrete frames is shown in Fig. 5.10b. Points A and B correspond to points of cracking and yielding, respectively. Line CD is parallel to OA and twice as long as OA. Similarly, DE is parallel to AB and twice the length of AB. A model is shown in Fig. 5.10c which allows for stiffness degradation due to load reversal in reinforced concrete frames that yield in flexure. Many models of this type are available in the literature [5.9, 5.10]. In Fig. 5.10c, line BC is parallel to OA. From C the stiffness changes and the line heads towards point D, the yield point in the negative direction. The line EF is parallel to line OA, but at F the slope changes, and the line heads towards point B, the load reversal point in the previous cycle.

The most popular method for non-linear analysis is the step-by-step direct integration technique, in which the time domain is discretized into many small intervals of time Δt, and for each interval with some assumption on the variation of acceleration, the equation of motion is solved using the displacement, velocity and acceleration of previous time step. The stiffness is assumed to be constant during each time step. Often the tangent stiffness at the beginning of the time step given by line AC of Fig. 5.11 is taken to be constant in a step-by-step manner as described below.

Consider the SDOF in Fig. 5.3, with the force–displacement hysteresis similar to that in Fig. 5.10. If (Eq. 5.1) is written for time t_i and $t_i + \Delta t$, then the incremental equation of motion, according to Eq. (5.2), takes the form

$$m \, \Delta \ddot{v}_i(t) + C \, \Delta \dot{v}_i(t) + k_i \, \Delta v_i(t) = \Delta P_i(t) \tag{5.29}$$

where

$$\Delta P_i(t) = -m\{\ddot{v}_g(t_i + \Delta t) - \ddot{v}_g(t_i)\} \tag{5.30}$$

where Δv_i, $\Delta \dot{v}_i$ and $\Delta \ddot{v}_i$ are the incremental displacement, velocity and acceleration of the mass, and ΔP_i is the incremental force. Furthermore,

$$k_i = \left(\frac{dF}{dv}\right)_{v = v_i} \tag{5.31}$$

which is the tangential slope (Fig. 5.11). Among the procedures available for

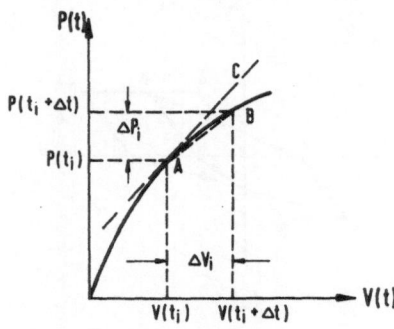

Fig. 5.11. Determination of stiffness for step-by-step integration technique.

performing step-by-step integration of Eq. (5.29), two popular methods are:

1. the constant acceleration method, where the acceleration is assumed to be constant over the interval, t_i and $t_i + \Delta t$

2. the linear acceleration method, where the acceleration is assumed to be linear over the interval t_i and $t_i + \Delta t$.

If the acceleration is assumed to vary linearly, then velocity and displacement must vary in a parabolic and cubic polynomial form. The increments in velocity and displacement become

$$\Delta \dot{v}_i = \ddot{v}_i \, \Delta t + \tfrac{1}{2} \Delta \ddot{v}_i \, \Delta t$$

$$\Delta v_i = \dot{v}_i \, \Delta t + \tfrac{1}{2} \ddot{v}_i \, \Delta t^2 + \tfrac{1}{6} \Delta \ddot{v}_i \, \Delta t^2 \tag{5.32}$$

Solving Eq. (5.32) for $\Delta \ddot{v}_i$ and $\Delta \dot{v}_i$ and substituting the expression into Eq. (5.29) leads to

$$\Delta v_i = \Delta \bar{P}_i / \bar{k}_i \tag{5.33}$$

where

$$\bar{k}_i = k_i + \frac{6m}{\Delta t^2} + \frac{3c}{\Delta t}$$

$$\Delta \bar{P}_i = \Delta P_i + m \left\{ \frac{6}{\Delta t} \dot{v}_i + 3\ddot{v}_i \right\} + c \left\{ 3\dot{v}_i + \frac{\Delta t}{2} \ddot{v}_i \right\} \tag{5.34}$$

$\Delta \dot{v}_i$ is now obtained, using Δv_i from Eq. (5.32). Thus, at the end of the time step

$$v_{i+1} = v_i + \Delta v_i$$

$$\dot{v}_{i+1} = \dot{v}_i + \Delta \dot{v}_i \tag{5.35}$$

Finally the acceleration $\ddot{v}_{i+1}$ at the end of the time step is obtained directly from the equation of motion as

$$\ddot{v}_{i+1} = \frac{1}{m} \{ P_{i+1} - c\dot{v}_{i+1} - k_{i+1}v_{i+1} \} \tag{5.36}$$

where the stiffness is evaluated at time t_{i+1} $(=t_i + \Delta t)$. The procedure is repeated for the next time step. The errors introduced on the assumption of linear variation of acceleration and constant stiffness during the time step are small if the time step Δt is short. However these errors generally tend to accumulate. This accumulation of error is avoided by imposing a total dynamic equilibrium condition at each step, as given by Eq. (5.36).

The accuracy of the results increases if smaller time steps are used. The following factors should be taken into account in selecting the time step:

1. the natural period of the structure – a time interval smaller than one-tenth of the period – is recommended

2. the rate of variation of the loading function – the time interval should be small enough to represent properly the variation of load with time

3. the complexity of the stiffness function – it is desirable to use smaller steps in the neighbourhood of drastic changes.

The above numerical procedures can be easily extended to MDOF systems. When the linear acceleration method is used for MDOF systems, the acceleration is assumed to be linear over the time interval t_i to $t_i + \theta \, \Delta t$, where $\theta \geq 1.38$ for the solution to be unconditionally stable. This procedure is called the Wilson-θ method [5.11].

References

5.1. Goodman L E, Rosenblueth E and Newmark N M, Aseismic design of firmly founded elastic structures, *Transactions ASCE* 1955; 120: 782–802.

5.2. Wilson E L, Der Kiureghian A and Bayo E P, A replacement for the SRSS method in seismic analysis, *International Journal of Earthquake Engineering and Structural Dynamics* 1981; 9: 187–194.

5.3. *NBC: National Building Code of Canada*, National Research Council of Canada, Ottawa, Canada, 1985.

5.4. Seed H B, Ugas H and Lysmer J, Site dependent spectra for earthquake resistant design, *Bulletin of the Seismological Society of America* 1976; 66: 221–243.

5.5. Balendra T and Heidebrecht A C, A foundation factor for earthquake design using Canadian Code of Practice, *Canadian Journal of Civil Engineering* 1987; 14: 498–509.

5.6. Applied Technology Council, *Tentative Provisions for the Development of Seismic Regulations for Buildings*, ATC 3–06, National Bureau of Standards, SP 510, 1978.

5.7. Balendra T, Tan T S and Lee S L, An analytical model for far-field response spectra with soil amplification effects, *Journal of Engineering Structures*, 1990; 12: 263–268.

5.8. Smith J W, *Vibration of Structures: Application in Civil Engineering Design*, Chapman and Hall, London/New York, 1988.

5.9. Clough R W and Johnston S B, Effect of stiffness degradation on earthquake ductility requirements, *Proceedings of the Second Japan Earthquake Engineering Symposium*, Tokyo, 1966, pp 227–232.

5.10. Takeda T, Sozen M A and Nielson N N, Reinforced concrete response to simulated earthquakes, *Journal of Structural Division*, ASCE 1970: 2557–2573.

5.11. Clough R W and Penzien J, *Dynamics of Structures*, McGraw-Hill, New York, 1975.

Chapter 6

Earthquake-resistant Design of Buildings

6.1 Design Philosophy

The current seismic design philosophy emphasizes the safety of lives in the event of a severe earthquake. As the nature and occurrence of earthquakes are indeterminate, it is necessary to consider different levels of earthquake intensity in the design of earthquake-resistant structures. This requirement is expressed in three levels of structural performance as follows:

1. The structure is to resist minor earthquakes without any damage.
2. The structure should resist moderate and frequently occurring earthquakes without any structural damage, but limited non-structural damage may be tolerated.
3. The structure should not collapse under rare and severe earthquakes.

Exceptions to these requirements include essential facilities such as hospitals, where more stringent criteria must be followed.

The provisions permit structures to yield under strong earthquake loadings, since it is generally uneconomical to design structures that remain elastic under severe earthquakes which have a low probability of occurrence during the life-span of the structures. The ability of structures to sustain loads in the plastic and strain hardening regimes provides an economical solution for seismic design under severe earthquake loadings.

As a consequence of the above seismic design philosophy, the design of seismic-resistant buildings in regions of high seismic activity must satisfy two general criteria. First, under frequently occurring low-to-moderate earthquakes, the structure should have sufficient strength and stiffness to control deflection and prevent any structural damage. Second, under rare and severe earthquakes, the structure must have sufficient ductility to prevent collapse. A minimum functional condition of the building should be maintained so that any trapped occupants may safely escape after the earthquake. Thus, in order to resist severe earthquake, the structural systems of buildings must have sufficient ductility in addition to stiffness and strength. The ductility is obtained through the

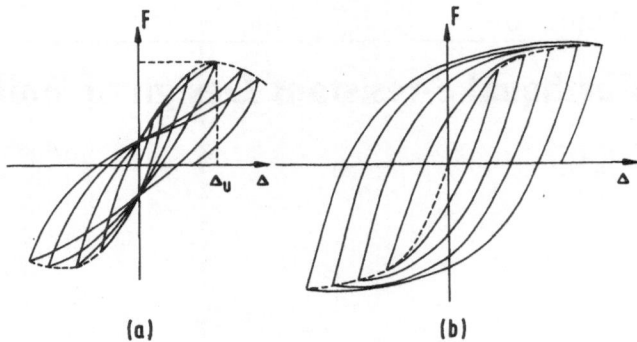

Fig. 6.1. Force displacement history under repeated loading: (a) poor in ductility, (b) good in ductility.

formation of stable plastic hinges which provide reasonably constant high levels of lateral load resistance during several reversed cyclic inelastic displacements, with as little loss of stiffness as possible after every such cycle. Figures 6.1a and 6.1b show the load–displacement hysteresis of loops of two different structures. In the first case (Fig. 6.1a) the earthquake-resistant capacity is poor since the strength and stiffness deteriorate under repeated loading and the hysteresis loops are pinched, which reduces the energy dissipation capacity, measured by the area under the hysteresis loop. In the second case (Fig. 6.1b) the earthquake-resistant capacity is good since the structure exhibits stable hysteresis loops without strength or stiffness degradation, and thus the energy dissipation is large.

6.2 Structural Configuration

The structural systems discussed in Chapter 2 can be made ductile through proper design and detailing. Both steel and concrete have been used successfully in designing ductile systems for earthquake-resistant structures. However, there are some important considerations with regard to vertical and horizontal configurations, as outlined below.

6.2.1 Vertical Configuration

Soft First Storey

When the stiffness and strength of the first storey is significantly less than those of the storeys immediately above, the deflection at first-storey level will be large, and plastic deformation tends to concentrate in the first storey which may cause the entire building to collapse as a result of failure of the first-storey columns. Soft first storey could occur:

(a) if the heights of first-storey columns are much higher than those of columns in the other storeys
(b) if the shear walls are made discontinuous below the second storey.

The soft storey is avoided by fitting additional columns or bracing in the soft storey, which will increase the stiffness and strength to a level comparable with the other storeys.

Short Columns

If both short and long columns exist in the same storey, then the relatively stiff column will attract a larger portion of the storey shear and could fail. This problem could arise at sites on hilly ground or by the introduction of a mazzanine floor which stiffens some of the columns in a particular storey, leaving the rest of the columns at their full unbraced length. This problem can be avoided by introducing spandrels which convert the long columns into short columns.

Vertical Setbacks

A vertical setback is a horizontal offset in the plan of the exterior boundary of a structure. Setbacks are introduced when a smaller floor area is required at upper levels or to admit light into adjoining sites. Asymmetrical setback introduces torsional forces, resulting in complex behaviour. Thus a symmetrical setback is preferable.

6.2.2 Horizontal Configuration

Simple symmetrical configurations such as square or circular shapes are preferable. In buildings with wings such as L, T, H, +, etc., damage will occur at the intersection of the wings. This problem can be avoided by separating the building structurally into simple shapes by means of seismic joints. There must be enough clearance at the seismic joints so that the adjoining portions do not pound each other.

In an asymmetrical layout, the centre of stiffness and the centre of mass do not coincide and thus the building twists, with the corner columns being severely punished. In order to improve torsional rigidity, some of the vertical resisting elements should be placed further away from the centre.

6.3 Steel Structures

As steel exhibits a high level of material ductility and energy absorption, it is an ideal material for earthquake-resistant structures. However, in order to take advantage of the inherent ductility of steel, considerable care is needed in the design and detailing of the members and connections of the structural systems. Commonly used structural systems in steel buildings are:

(1) the moment-resisting frame (MRF)

(2) the concentric braced frame (CBF).

MRFs are ductile and thus excellent for energy dissipation. However, they tend to be flexible to control the drift unless the sizes of the beams are increased. On the other hand, CBFs are stiff but suffer from lack of ductility owing to buckling of the braces. Thus neither MRFs nor CBFs provide an economical solution to seismic-resistant design. However, it is feasible to obtain an economical system which combines the advantages of MRFs with those of CBFs. Two such systems are:

(1) the eccentric braced frame (EBF)

(2) the knee-brace-frame (KBF).

These two systems strike a balance between stiffness, strength and ductility to provide an economical solution. The design principles of these structural systems are presented below.

6.3.1 Moment-Resisting Frame (MRF)

The MRF derives its stiffness, strength and ductility from the flexural resistance of the beams and columns. It is customary to design MRFs on the basis of the strong column–weak beam concept. The columns are designed to remain elastic, except for bottom-storey columns where plastic hinges may be present in an approved manner, as shown in Fig. 6.2.

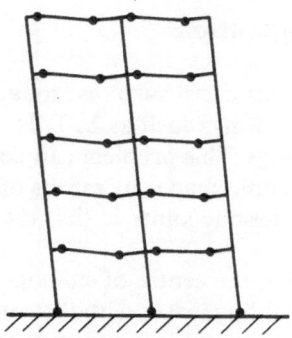

Fig. 6.2. Failure mode for a typical multistorey frame with plastic hinges at column ends.

Design of Beams

The beams are designed to yield with adequate plastic rotation capacity so that they provide energy absorption. For stable hysteresis behaviour under cyclic loading, which is desirable for energy absorption, local and lateral torsional buckling of the beams must be avoided or controlled. Lateral torsional buckling is controlled by providing lateral restraints to the compression flanges at particular spacings, as stipulated in Table 6.1.

Local buckling of webs or flanges of a beam is less serious because of the significant post-buckling strength of plate elements. However, such buckling will lead to pinching of the hysteresis loops and hence reduces the energy dissipation, which is undesirable. There is also a tendency for the magnitude of the buckles to grow with each successive cycle, and the local distortion of the section may increase the lateral deformation elsewhere and thereby reduce the overall stability of the beam. Thus to prevent local buckling, the maximum width to thickness ratios must satisfy the values given in Table 6.2.

Design of Columns

When plastic hinges are allowed to occur in columns, the ability for the columns to dissipate energy in a stable flexural mode depends on the ability against lateral–torsional buckling and local buckling. The level of axial load is also important as it can increase the tendency for lateral or local buckling, leading to accelerated strength degradation. To control lateral–torsional buckling, Butterworth and Spring [6.2] recommend lateral restraints to the flanges in the plastic hinge zone L_y, defined as the length over which the moment, M, exceeds 75% of the plastic moment of the column M_{pc}, allowing for the effects of axial compression load. The spacing of the lateral restraints is the same as for the beams given in Table 6.1, except for the definition of L_y. However, if the axial load, P, is less than 15% of the squash load P_y ($= A_s f_y$), where A_s is the cross-sectional area and f_y is the yield stress, then L_y is the same as for beams. To prevent local buckling, the width to thickness ratios must comply with the limits given in Table 6.3.

Table 6.1. Spacing of lateral restraints for fully ductile members [6.1].

	$L_y > 480r_y/f_y^{1/2}$	$L_y \leq 480r_y/f_y^{1/2}$
Spacing of braces within length L_y	$\leq 480r_y/f_y^{1/2}$	one brace required
Spacing of brace adjacent to length L_y	$\leq 720r_y/f_y^{1/2}$	$\leq 720r_y/f_y^{1/2}$

L_y = Flange length where moment M exceeds 85% of the plastic moment of resistance of the beam.
r_y = Radius of gyration about the minor axis.
f_y = Yield stress in N/mm².

Table 6.2. Maximum width to thickness ratios for fully ductile members [6.1].

Flanges and plates in compression with one unstiffened edge (e.g. flanges of I or [sections)	$\dfrac{b_1 f_y^{1/2}}{t_f}$	120
Flanges of welded box sections in compression	$\dfrac{b_2 f_y^{1/2}}{t_f}$	500
Flanges of rectangular hollow sections	$\dfrac{b_2 f_y^{1/2}}{t_f}$	350
Webs under flexural compression	$\dfrac{d_1 f_y^{1/2}}{t_w}$	1000
Webs under uniform compression	$\dfrac{d_1 f_y^{1/2}}{t_w}$	500

Notes
(1) The symbols b_1, b_2, d_1, t_f and t_w are defined in Fig. 6.3.
(2) f_y is the yield stress in N/mm^2.

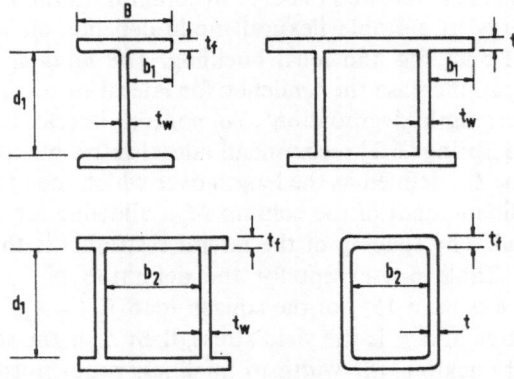

Fig. 6.3. Section nomenclature for Table 6.2.

The column must be checked for its load-carrying capacity, taking into account the moment–axial load interaction. For I sections, using a linear equation for moment–axial load interaction, the load-carrying capacity of the columns at the support section is given by [6.2]:

(i) Bending about the major principal axis

$$\frac{M}{M_p} \le 1.0 \text{ for } \frac{P}{P_y} < 0.15 \tag{6.1}$$

Table 6.3. Maximum width to thickness ratios for fully ductile columns [6.2].

$\dfrac{b}{t_f}$ (I column flange)	$\leq 120/f_y^{1/2}$
$\dfrac{b}{t_f}$ (box flange)	$\leq 500/f_y^{1/2}$
$\dfrac{d}{t_w}$ (web)	$\leq 500/f_y^{1/2}$

f_y = Yield stress in N/mm^2.
t_f = Flange thickness.
t_w = Web thickness.
d = Clear depth of column web.
b = Outstand of flange beyond connection
      to web (I columns), or clear distance
      between sides of box columns.

$$\frac{P}{P_y} + 0.85 \frac{M}{M_p} \leq 1.0 \text{ for } \frac{P}{P_y} \geq 0.15 \tag{6.2}$$

(ii) Bending about the minor principal axis

$$\frac{M}{M_p} \leq 1.0 \text{ for } \frac{P}{P_y} < 0.4 \tag{6.3}$$

$$\left(\frac{P}{P_y}\right)^2 + 0.85 \frac{M}{M_p} \leq 1.0 \text{ for } \frac{P}{P_y} \geq 0.4 \tag{6.4}$$

(iii) Bending about both principal axes

$$\frac{M_x}{M_{px}} + \frac{M_y}{M_{py}} \leq 1.0 \text{ for } \frac{P}{P_y} < 0.15 \tag{6.5}$$

$$\frac{P}{P_y} + 0.85 \frac{M_x}{M_{px}} + 0.85 \frac{M_y}{M_{py}} \leq 1.0 \text{ for } \frac{P}{P_y} \geq 0.15 \tag{6.6}$$

where M_x, M_y and M_{px}, M_{py} denote the moment and plastic moment about the x and y axes.

Design of Connections

The moment-resisting beam–column connections of MRFs must accommodate the inelastic response of the members. The normal practice is for the connection to remain elastic. Bolted connections may lead to pinched hysteresis loops as a result of slippage. Thus all welded or bolted web and welded flange

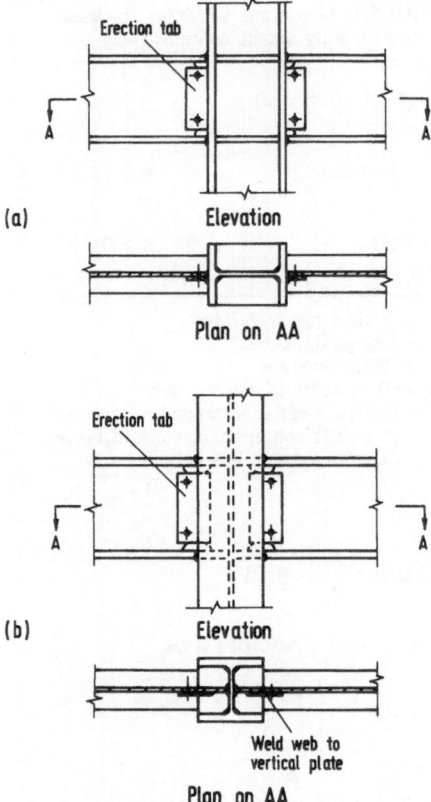

Fig. 6.4. Welded beam–column connection: (a) about strong axis, (b) about weak axis.

connections are preferred to ensure sustainable ductility of the frame. Typical welded beam–column connections are shown in Fig. 6.4.

The forces acting on a beam–column connection are shown in Fig. 6.5. The forces to be resisted are those arising when plastic hinges are formed in the connection members; the variability of yield strength above the minimum value and the over-strength due to strain hardening should be taken into account. The typical deformation of the connection under failure load is shown in Fig. 6.6. As shown in the figure, at the location of compression beam flanges, the column web could cripple and at the locations of the tension beam flanges, the column flange weld could fracture.

The crippling of column web due to compression from the beam flange can be avoided by ensuring

$$\frac{d_c}{t_{cw}} \le \frac{473}{\sqrt{f_{yc}}} \tag{6.7}$$

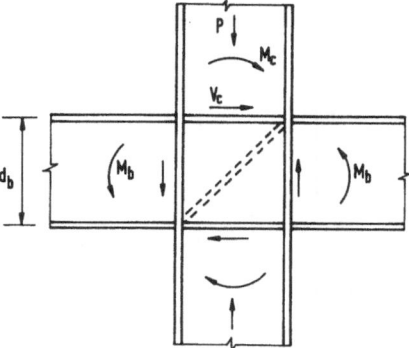

Fig. 6.5. Forces acting on beam–column connection.

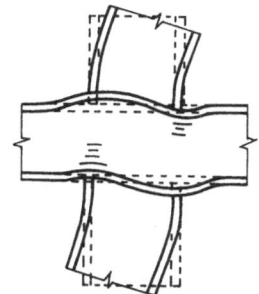

Fig. 6.6. Deformation of beam–column connection.

where d_c and t_{cw} are the depth and thickness of the column web and f_{yc} is the yield stress in N/mm^2.

To prevent web yielding under compressive load from the beam flange, the thickness of the column web must satisfy the following expression [6.3]:

$$t_{cw} \geq \frac{1.5B_b T_b}{T_b + 5k + 2T_{ep} + 2w_f} \frac{f_{yb}}{f_{yc}} \tag{6.8}$$

where f_{yb} is the beam yield stress and f_{yc} the column yield stress. The other symbols are defined in Fig. 6.7.

The flexural failure of the column flange due to tensile force from the beam flange can be avoided if [6.4]

$$t_{cf} > 0.6\left(B_b T_b \frac{f_{yb}}{f_{yc}} \right)^{1/2} \tag{6.9}$$

where t_{cf} is the thickness of the column flange, B_b and T_b are the breadth and thickness of the beam flange, f_{yb} is the beam yield stress and f_{yc} is the column yield stress.

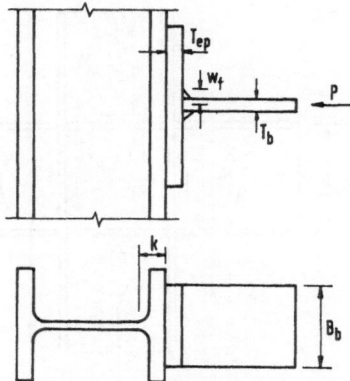

Fig. 6.7. Parameters affecting the dispersion of load from the compression beam flange.

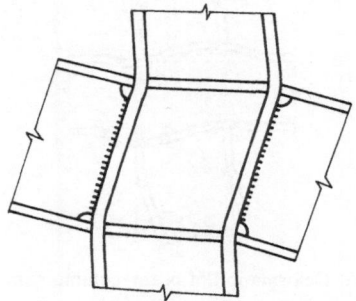

Fig. 6.8. Shear yielding of panel zone.

The panel zone of the connection may yield under shear, as shown in Fig. 6.8. To prevent this mode of failure, the shear force acting in the panel zone, V, must be less than the shear capacity of the panel V_{py}. Based on the von Mises yield criterion

$$V_{py} = 0.55 f_{yc} t_{cw} d_c \qquad (6.10)$$

and from equilibrium of forces in Fig. 6.5

$$V = \frac{2M_b}{d_b} - V_c \qquad (6.11)$$

where V_c is the shear in the column, M_b the moment in the column, d_b the depth of the beam, d_c the depth of the column, t_{cw} the thickness of the column web and f_{yc} the yield stress of the column web.

In order to satisfy the above criterion, doubler plates or diagonal stiffeners may be used in the panel zone.

6.3.2 Concentrically Braced Frame (CBF)

Concentrically braced frames are commonly used to resist wind forces. The diagonal bracing provides an efficient and economical way of resisting lateral loads. Because of buckling of the braces, CBFs have low system ductility. However, as CBFs tend to be stiffer than MRFs, a combination of these two systems is commonly used to resist seismic loads. Typical arrangements of braces in CBFs are shown in Fig. 6.9. For X bracing or Z bracing, the axes of the braces pass through the beam–column joints, whereas for V bracing the pair of braces meet at a point along the beam. In the case of Z bracing, the braces in different bays slope in opposite directions in order to avoid the large residual sway deflection that would occur if the braces were arranged asymmetrically.

If the braces are very slender and hence capable of tensile resistance only, there will be undesirable impact type response as the braces straighten out from the buckled zero load state to the tensile load-carrying state. Thus it is good practice to provide braces that are capable of compressive as well as tensile resistance.

The local, lateral–torsional buckling and tensile yielding of the braces reduce the stiffness, strength and energy dissipation capacity of CBFs under cyclic loading. The deflection ductility factors μ [6.5] that can be achieved by CBFs having negligible bending strength are given in Table 6.4. The value of μ decreases with increase in slenderness ratio of the brace. It also decreases with number of storeys, and not more than three storeys are recommended with this type of construction. However, if the frame connections are made moment

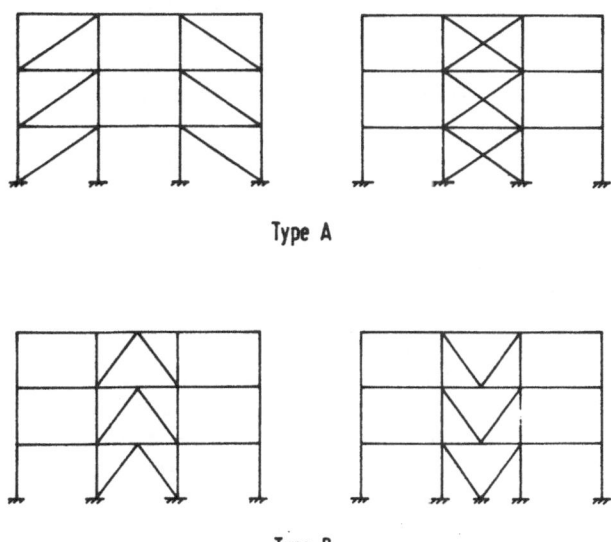

Type A

Type B

Fig. 6.9. Types of concentric braced frame.

Table 6.4. Design values of μ for CBFs having negligible bending strength [6.5].

Bracing type (Fig. 6.9)	No. of storeys	Bracing slenderness $\dfrac{Kl}{r}\sqrt{\left(\dfrac{f_y}{250}\right)}$		
		< 40	41–48	81–135
		μ	μ	μ
Type A	1	5.0	3.9	3.0
	2	4.5	3.5	2.4
	3	4.0	3.0	1.8
Type B	1	3.7	2.4	1.5
	2	3.0	1.8	1.2
	3	2.7	1.5	1.0

resisting and with appropriate energy dissipating devices in the brace, CBFs can be used for buildings with more than three storeys.

6.3.3 Eccentrically Braced Frame (EBF)

In an eccentrically braced frame, the centre line of the brace is eccentric to the beam–column joint, as shown in Fig. 6.10. The short segment of the beam between the brace–beam joint and the beam–column joint is called the *active link*. The axial force in the brace is transmitted to the column through shear and bending in the link. Under severe earthquake this link yields in shear to dissipate energy and prevent brace buckling. Thus in the alternative types of EBFs shown in Fig. 6.10, each brace is connected to at least one link which can undergo large displacements to prevent brace buckling. Unlike CBFs, EBFs can accommodate such architectural features as door and window openings with less interference.

Elastic Behaviour

The influence of the eccentricity e, the length of the active link, on the elastic stiffness is shown in Fig. 6.11 for different aspect ratios of the frame [6.6] and a fixed set of sectional properties. As e/L varies from 0 to 1.0, the system changes from CBF to MRF. It is seen that for $e/L > 0.5$, there is no benefit from the brace. However, as e/L values decrease there is a substantial increase in stiffness. Thus for higher stiffness, smaller values of e/L are preferred.

For frames with wide bays, a single brace cannot be used since the length of the brace becomes too long. In this case, split K bracing as shown in Fig. 6.11b

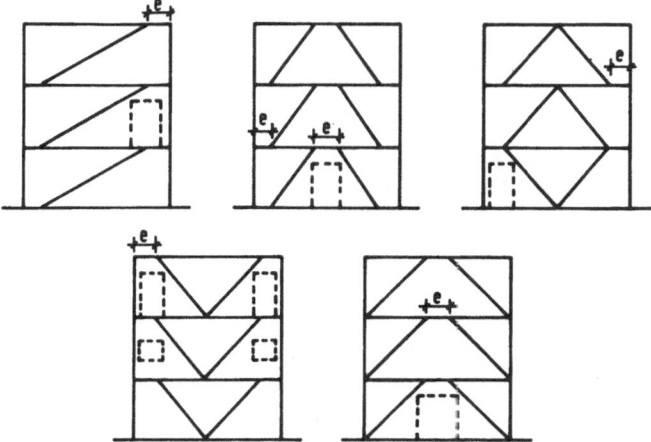

Fig. 6.10. Types of eccentric braced frame.

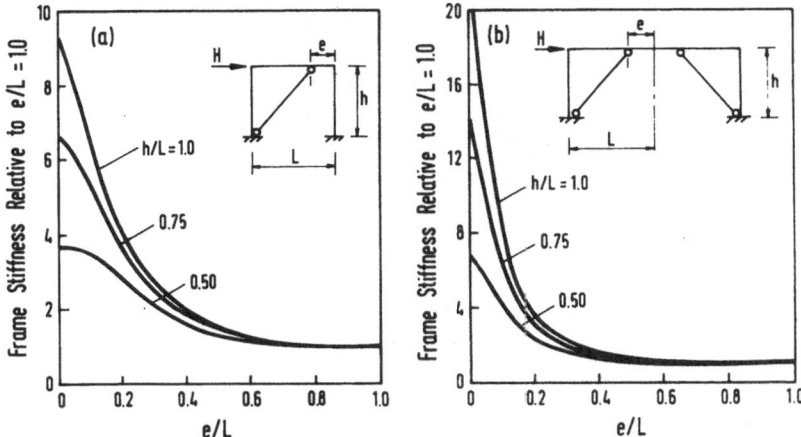

Fig. 6.11. Influence of eccentricity on the elastic stiffness of EBFs.

could be used. The stiffness of this framing system is more sensitive to e/L ratio than the one with a single brace.

Inelastic Behaviour

Because of larger shear forces, the link will yield in shear when it is short. Figure 6.12 shows the collapse mechanism using rigid plastic theory. From simple geometry, the frame deformation θ is related to member deformation γ by

$$\theta L = \gamma e \text{ or } \gamma = \theta/(e/L) \tag{6.12}$$

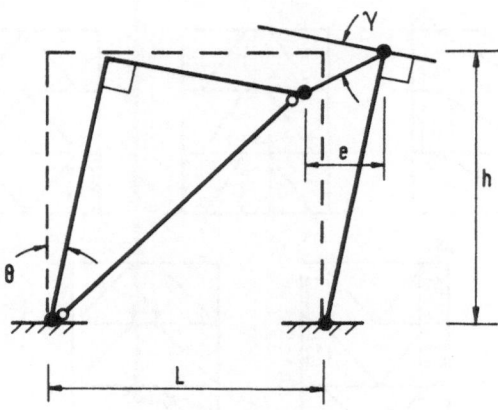

Fig. 6.12. Collapse mechanism of one-storey EBFs.

Thus, the smaller the eccentricity, the larger will be the deformation of the link or the ductility demand. Although a shorter link is preferred for stiffness, in order to meet the ductility demand, the length of the link will be governed by the available ductility in the link. Experimental testing shows that short links which yield in shear (shear links), when provided with suitable stiffeners to prevent web buckling, dissipate much more energy than longer ones with moment hinges (moment links). A shear link is created by ensuring that the webs of the link yield in shear. For this, the length of the link [6.7]

$$e \leq 1.6 M_p^* / V_p^* \tag{6.13}$$

where

$$M_p^* = f_y (d - t_f)(b - t_w) t_f \tag{6.14a}$$

$$V_p^* = \frac{f_y}{\sqrt{3}} (d - t_f) t_w \tag{6.14b}$$

where f_y is the tensile yield stress, d and b the depth and breadth of the WF beam, t_f the thickness of the flange and t_w the thickness of the web.

To prevent web buckling in the shear links under cyclic loadings, web stiffeners are required. The stiffeners should be placed $25 t_w$ to $30 t_w$ apart. It is sufficient to provide stiffeners only on one side of the web and they need not be welded to the top flange of the beam. The lateral restraints to prevent lateral buckling of the shear link are provided by the transverse floor beams. Test results indicate that properly stiffened shear links can sustain plastic shear rotations of up to ± 0.1 radian.

Design of Other Components of EBFs

For failure mode control, other members of EBFs need to be sufficiently stronger than shear links. For this, when estimating the ultimate capacity of the shear links, potential over-strength of the links due to strain hardening and variability in yield stress beyond the normal range should be considered.

As bolted beam web–column connections are prone to slippage, the link–column connection should be all welded. The link to brace connections are subjected to severe bending and thus require special consideration in the design.

Both full scale and scaled models of EBFs have been tested dynamically [6.8–6.10]. The test results show the excellent performance of EBFs in exhibiting high ductility and energy dissipation capacity. However, a significant floor damage could occur owing to the large shear deformation of the links.

6.3.4 Knee-Brace-Frame (KBF)

In a knee-brace-frame (KBF), one end of the diagonal brace is connected to a knee anchor instead of the beam–column joint, as shown in Fig. 6.13. In this system the diagonal brace provides the required stiffness during moderate earthquakes, while the knee anchor yields in flexure to dissipate energy in the event of a severe earthquake, and hence prevents brace buckling or yielding of beams or columns.

Elastic Stiffness

Through dimensional analysis, the elastic stiffness of KBFs may be expressed as follows:

$$\frac{K}{EI_c/H^3} = f\left\{\frac{I_k}{I_c}, \frac{A/l}{I_c/H^3}, \frac{b}{h}, \frac{h}{H}, \frac{H}{B}\right\} \tag{6.15}$$

where E is Young's modulus of elasticity, A the cross-sectional area of the brace, l the length of the brace and I_c, I_k the second moment of area of column

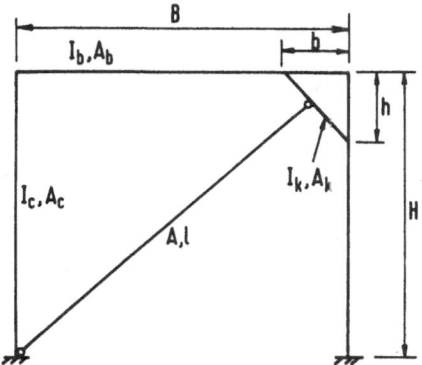

Fig. 6.13. Knee-brace-frame.

and knee, respectively. The parameter H/B represents the aspect ratio of the frame, while b/h and h/H describe the orientation and length of the knee anchor. The influence of various parameters on the stiffness of alternative types of KBF is depicted in Fig. 6.14. The stiffness increases substantially with the brace area. However, beyond a certain limit, any further increase in brace area produces only a small increase in stiffness. For a stiffer knee anchor, which corresponds to a larger value of I_k/I_c, the stiffness increases substantially. When the length of the knee is varied by changing the ratio h/H while keeping the orientation of the knee anchor (b/h) constant, higher stiffness is obtained for a shorter knee. The stiffness improves substantially for h/H values less than 0.3.

From Fig. 6.14 it is evident that KBFs with a knee anchor at the top or

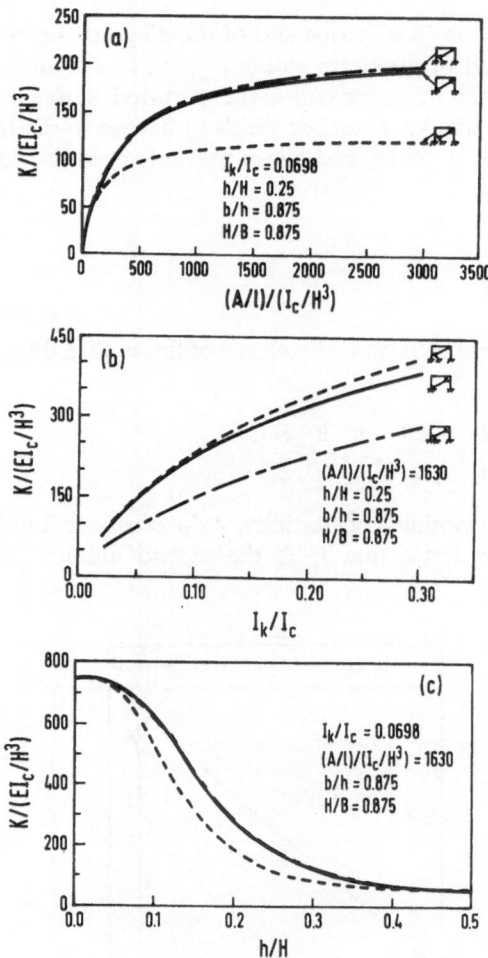

Fig. 6.14. Effects of: (a) brace area, (b) moment of inertia of knee anchor, (c) length of knee anchor, on the stiffness of KBFs.

bottom end of the brace are superior to those with a knee anchor at both ends of the brace.

Inelastic Behaviour of KBFs

The amount of energy that can be dissipated by a KBF depends on the ductility capacity of the knee anchor. In order to achieve sufficient ductility, local flange buckling and lateral–torsional buckling of the knee anchor must be eliminated or controlled. As lateral bracing of the knee anchor is generally difficult, a convenient solution is to use a square box section for the knee anchor. Because of its high torsional rigidity and the fact that its lateral stiffness is not less than its in-plane stiffness, there is no lateral–torsional buckling limit state for this section shape. Through pseudo-dynamic testing of large-scale KBFs, it has been found that hot rolled square box sections are suitable to give good ductile performance to knee anchors when they are designed against local buckling [6.11]. The results show that a well designed knee anchor can sustain a plastic rotation of ± 0.1 radian.

When a KBF is subjected to lateral loading, each segment of the knee anchor, namely segments between the beam-knee and knee-brace joints and the column-knee and knee-brace joints, deforms approximately in an antisymmetric manner. Based on this assumption, a bilinearized moment–rotation relationship can be established for each segment of the knee anchor [6.12]. The elastic stiffness (M/θ) would be $6EI/L$, where EI is the flexural rigidity and L the length of the segment. The yield moment M_y and the strain hardening coefficient β are determined from Fig. 6.15, where M_p is the plastic moment of the section and α is the strain hardening coefficient of the bjlinearized uniaxial stress–strain relationship. Using this moment–rotation relationship, the inelastic response of KBFs can be determined. Analytical studies conducted for

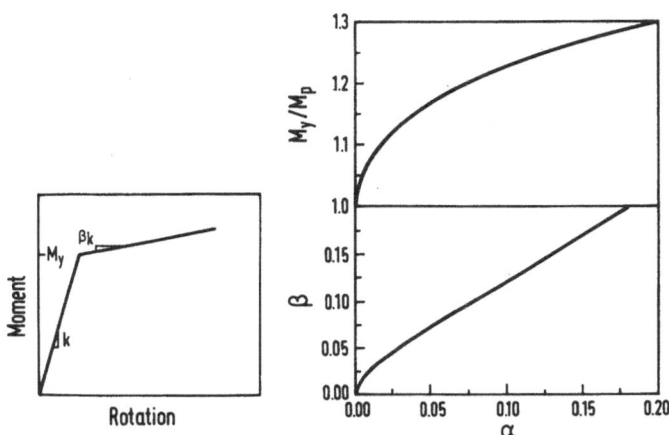

Fig. 6.15. Bilinear moment–rotation relationship for knee anchor.

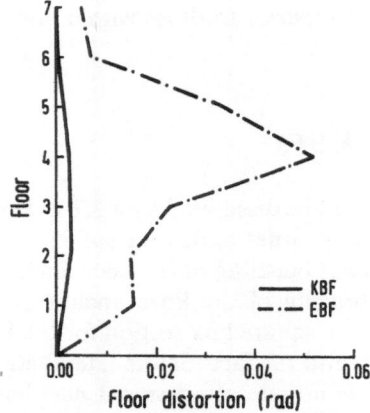

Fig. 6.16. Comparison of floor distortion envelopes of a seven-storey building using EBF and KBF systems.

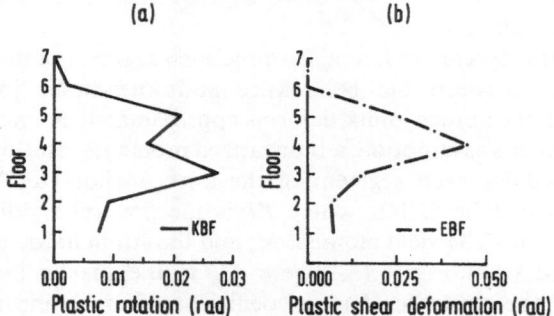

Fig. 6.17. Inelastic deformation of: (a) knee anchors, (b) shear links of a seven-storey building [6.13].

a seven-storey building [6.13] reveal that the KBF system is an attractive alternative to the EBF system. It has a clear advantage of greatly reduced floor deformation as illustrated in Fig. 6.16. The envelopes of inelastic deformation of the knee anchors and shear links of the building are depicted in Fig. 6.17. In the case of KBF the damage sustained during an extreme earthquake is confined to the knee anchor, which is a secondary member. Thus, a retrofitting process is much simpler and economical than in the case of EBF where the shear link is an integral part of the main structural member.

6.3.5 Applications

When adequate precautions are taken against premature local and lateral buckling of members as well as failure of connections, steel structures are very effective in resisting earthquakes owing to their lightness and inherent ductility

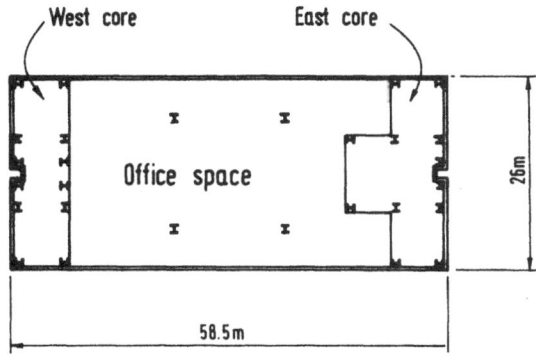

Fig. 6.18. Plan view of Bank of America [6.14]..

which provide sufficient reserves of strength at large displacements to absorb energy. As such, a substantial proportion of high-rise buildings in earthquake-prone regions are steel structures. In recent years, EBFs have become popular structural systems for seismic design. To illustrate the advantages of the EBF system, its application in the construction of the Bank of America in San Diego [6.14] and the Getty Plaza in Los Angeles [6.15] are discussed in this section.

The plan of the Bank of America is shown in Fig. 6.18. The building has a tower extending from a five-storey concrete base. The tower contains sixteen floors of office space with the first office level at 12.5 m above the plaza level. Thus the height of the tower is that of a 19-storey building. In plan the tower is 26 m wide and 58.5 m long. It consists of two cores about 6.5 m long at each end. The office area contains only four columns.

This building is located is seismic zone 3 as defined in the Uniform Building Code [6.16]. According to this code, any building over 50 m height must incorporate ductile moment-resisting frames to resist lateral loads. Either complete ductile frame systems or dual systems are permitted. The latter must consist of braced frames capable of resisting 100% of the code-specified lateral forces and a system of completely independent ductile frames capable of resisting not less than 25% of the code-specified lateral forces. For a complete ductile moment frame solution, the close column spacing in the core areas could perform efficiently as components in moment frames, however the wide column spacing in the large area between the cores could not be used efficiently in the moment-resisting frame owing to long beam spans. One could have moderate beam spans above the first floor by introducing additional columns supported by transfer girders at the first floor level. However, this arrangement would lead to a soft first storey. As the performances of buildings with soft first storeys have been found to be disastrous during past earthquakes, a complete moment frame solution was not economically feasible. With the given building layout, it was not possible to accommodate a dual system, that is, a ductile moment frame and a braced frame. Thus, the EBF system was found to be best, as it has the merits of both a ductile moment frame and a braced frame in a single

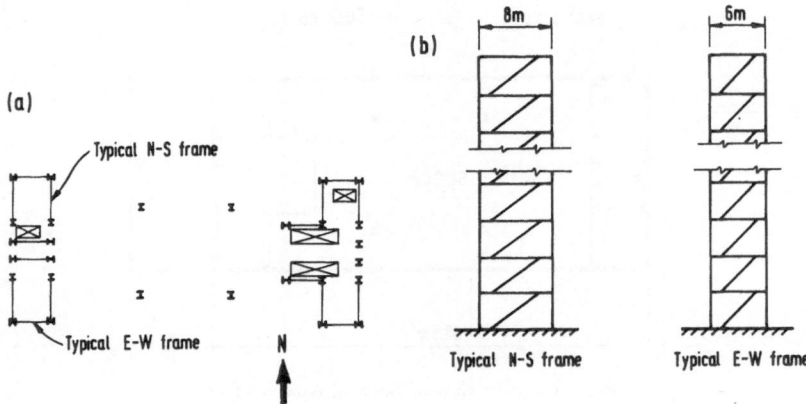

Fig. 6.19. Arrangements of EBFs: (a) in plan, (b) in elevation [6.14].

system. It can provide adequate stiffness to withstand moderate earthquakes, with small drift and ductility to absorb a great amount of energy during severe earthquakes.

The arrangements of EBFs are shown in Fig. 6.19. In each direction, eight EBFs with shear links at each end of the brace are provided. The details of all welded beam–brace connections are shown in Fig. 6.20. Structural tubes are used for the braces. Details of shear link to column flange and shear link to column web connections are shown in Fig. 6.21. All welded moment connections are used. Web stiffeners are provided to prevent web buckling of shear links. Since the shear link to column web connections are not as reliable as the shear link to column flange connections, the link to column web connections are placed at the lower ends of the braces, as links at lower ends are subjected to less deformation.

The 36-storey Getty Plaza is chosen as the second example to illustrate the effectiveness of the EBF system as one of the components of the dual system for seismic design. The typical floor plan of this building is shown in Fig. 6.22. If concentric braced frames were used for stiffness, their locations would be restricted to the elevator core or other service areas in order to prevent any interference with architectural planning. Since the widths of these areas were

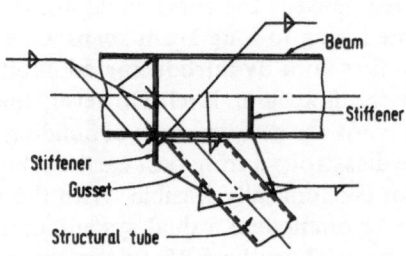

Fig. 6.20. Beam–brace connection.

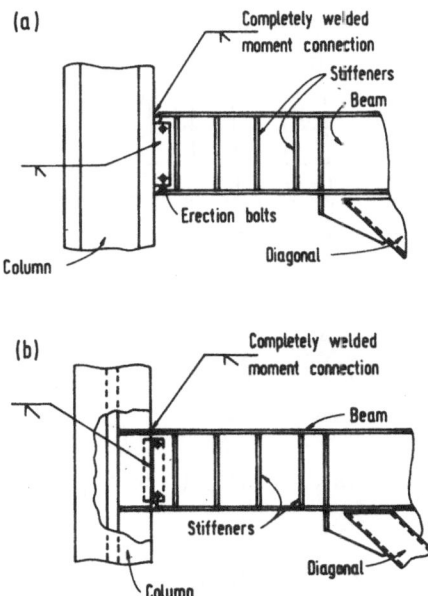

Fig. 6.21. Shear link to column connections: (a) shear link to column flange, (b) shear link to column web.

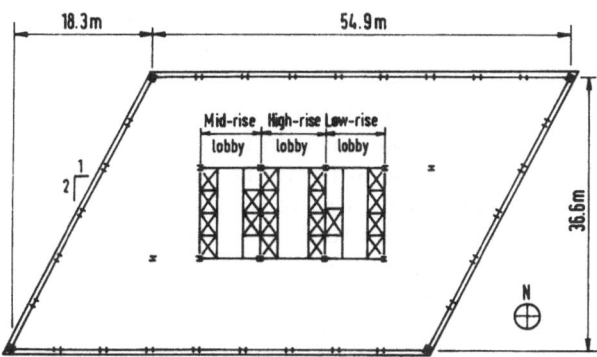

Fig. 6.22. Typical floor plan of Getty plaza [6.15].

small compared to the overall dimensions of the building, CBFs would need to be slender and their stiffness would not be adequate to control the drift. Thus additional framing such as a perimeter framed tube would be required to meet the serviceability requirements at moderate excitation. Furthermore, for severe excitation, as CBFs are not ductile they must be designed to resist 100% of the code-specified lateral loads while the independent ductile framed tube must be designed to resist 25% of the lateral loads. Thus the code penalizes

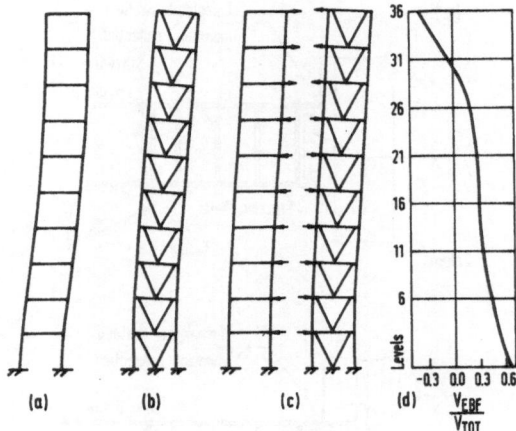

Fig. 6.23. Framed tube–EBF interaction [6.15]: (a) frame, (b) braced frame, (c) interaction behaviour, (d) shear in EBF.

CBFs by requiring them to carry the total seismic shear. The seismic shear cannot be shared between the framed tube and CBFs. On the other hand if EBFs are used instead of CBFs, as EBFs are ductile, the seismic base shear can be shared between EBFs and the framed tube. Using EBFs would not violate the stiffness requirement because of their inherent rigidity. Thus in this project the dual system comprising framed tube and EBFs were used. The EBFs were located between adjacent elevator cores. Because of the wide spacing of the columns, inverted K configuration was used with shear links at both ends of the beam adjacent to the columns. The shear in EBF was determined using the interaction between the framed tube which deforms in shear mode and the EBF which deforms in flexural mode. As seen from Fig. 6.23, EBFs are supported at the top by the framed tube, whereas the framed tube is supported by the EBFs at the bottom. The EBFs need to be designed to carry only about 60% of the total base shear instead of 100% as specified for the CBFs. Hence EBFs in conjunction with the framed tube provide an economical solution in meeting both the stiffness and ductility requirements. It should be noted that at the top, the total shear carried by the framed tube exceeds the applied storey shear because of interaction between the framed tube and the EBFs.

6.4 Concrete Structures

A properly designed, detailed and constructed reinforced concrete structure can withstand severe earthquakes because of its ductile behaviour. For a reinforced concrete section, the ductility capacity increases with

(a) increase in the content of compression steel

(b) increase in the compressive strength of the concrete

(c) increase in ultimate concrete strain,

and decreases with

(a) increase in the content of tensile steel

(b) increase in the yield strength of the steel

(c) increase of axial load.

The desired ductility capacity for a structure is achieved by adopting the following principles:

(a) under-reinforced sections are used in order to stabilize the ductility;

(b) beams are designed to yield while the columns remain elastic, so energy is dissipated through plastic moment hinges in the beams of framed structures;

(c) in coupled shear walls, plastic hinges are located in the coupling beams so that flexural cracking occurs at the base of the wall;

(d) brittle failure due to shear is avoided by ensuring that the shear capacity of the sections is larger than the shear at ultimate moment capacity;

(e) brittle failure due to inadequate bonding is prevented by avoiding high stressed areas of concrete for anchoring and splicing reinforcement bars;

(f) core concrete in beams and columns is confined by stirrups or spiral reinforcement, which increases the ultimate concrete strain in addition to providing lateral support for the reinforcement;

(g) concrete with a minimum strength of $20 \, N/mm^2$ and reinforcement with yield stress of not more than $410 \, N/mm^2$ are used.

Building codes, such as those of the American Concrete Institute, ACI Committee 388 [6.17] and the New Zealand Standard Code of Practice [6.18], provide recommendations on detailing the reinforced concrete elements for seismic loading. A brief review of the recommendations for the design of various elements is presented below. Further discussion on the subject is given by Park and Paulay [6.19].

6.4.1 Moment-resisting Frames

Beams

For beams to behave as compact elements and to ensure effective transfer of moments between beam and columns, the following restrictions are placed on the size of beams:

(a) b not less than 250 mm

(b) b/h not less than 0.3

(c) b not greater than the column width plus $0.75h$ on each side

(d) l/h not less than 4

where b is the breadth, h the depth and l the span of the beam.

The minimum bending moments in the beam are specified as follows.

1. the positive moment strength of the beam at the face of the column shall be not less than one-half of the negative moment strength;
2. neither the negative nor the positive moment strength at any point in the beam shall be less than one-quarter of the maximum moment at the face of the column.

The upper and lower limits on longitudinal reinforcement content, as a fraction of cross-sectional area of the beam ($h \times b$), are 0.25 and $1.4/f_y$, respectively, where f_y is the yield strength in N/mm^2. The minimum permissible size for the longitudinal bar is 12 mm and there must be at least two bars in both the top and the bottom of the section. The required amount of longitudinal reinforcement is calculated using ultimate strength theory [6.19].

Transverse reinforcement must be provided to resist shear. The spacing of the shear stirrups should not exceed $d/2$, where d is the effective depth. The minimum diameter for the stirrup is 10 mm.

The splices in the longitudinal reinforcement must be located in the zones of low stress. Splices are not acceptable in the plastic zone nor in the column zone. Where splices occur, the maximum spacing of the stirrups is $d/4$ or 100 mm.

Plastic Zone. Normally plastic hinges occur at the ends of the beam, and the plastic zone is assumed to extend from the column face to a distance of twice the beam depth. In order to provide adequate ductility in the plastic zone, the area of the compressional steel should be not less than 50% of the tensile steel. Furthermore, the tensile reinforcement ratio, ρ, should not exceed values given by [6.18]:

$$\rho_{max} = \frac{1 + 0.17(f'_c/7 - 3)}{100}\left(1 + \frac{\rho'}{\rho}\right), \quad \text{or} \tag{6.16}$$

$$\rho_{max} = 7/f_y$$

where ρ' is the ratio of compressional reinforcements and f'_c the concrete compressive strength.

A minimum of two 16 mm bars should be provided in both the top and bottom faces in the plastic zone. In order to ensure confinement of the concrete, the stirrup spacing in the plastic zone should not exceed $d/4$, six times the diameter of the largest main bar or 150 mm. The first stirrup tie must be within 150 mm of the column face.

A typical arrangement of reinforcement in beams is shown in Fig. 6.24.

Columns

To ensure plastic hinges cannot form in the columns, the ultimate moment strength of the columns should be higher than that of beams framing into

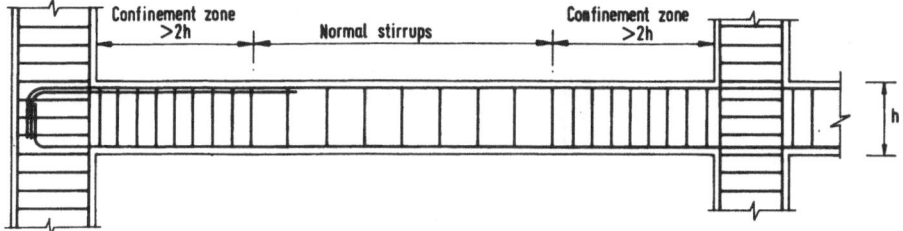

Fig. 6.24. Typical arrangement of beam reinforcements.

the columns along each principal plane of the structure. However, if seismic loading occurs in a general direction, yielding can occur in the column.

The recommended minimum size of the column is 300 mm × 300 mm [6.17]. The minimum content of the longitudinal steel should be 1% of cross sectional area. The maximum content should not be more than 6% for grade 300 steel ($f_y = 300$ N/mm^2). The percentage could be increased to 8% at the locations of splices, which are limited to the middle half of the column. The corresponding percentages for grade 400 steel ($f_y = 400$ N/mm^2) are 4.5% and 6%, respectively.

The minimum size of the longitudinal bar is 12 mm. In the plastic zone, the longitudinal bars should not be spaced more than 200 mm apart, and the smallest bar diameter should not be less than two-thirds of the largest bar diameter in any one row.

In the potential plastic zones, when spiral or circular hoops are used for confinement of concrete, the ratio between the volume of confining steel and volume of core, denoted as ρ_s, is given by

$$\rho_s = 0.12f'_c/f_{yh} \tag{6.17}$$

where f'_c is the concrete compressive strength and f_{yh} the yield strength of the stirrups. The total cross-sectional area, however, must not be less than a specified minimum value [6.17]. The spacing of the stirrups in the vertical direction should not be greater than one-quarter of the least column dimension or 100 mm. In plan, the spacing of the stirrup legs or cross ties should not exceed 350 mm. The minimum sizes for the stirrups and cross ties are 10 mm and 8 mm, respectively. The distance over which the confining steel is required is given by the greater of column depth, one-sixth of the clear column height or 450 mm.

Transverse reinforcements are provided in the form of multiple stirrups or stirrups and cross ties, positively anchored in order to restrain the longitudinal column bars and to give satisfactory confinement to the concrete. Spiral reinforcement is almost twice as efficient as rectangular hoops. To prevent unwinding of the spiral reinforcement, in the event of spalling of the concrete, it must be positively anchored. Typical reinforcement details of columns are shown in Figs. 6.25 and 6.26.

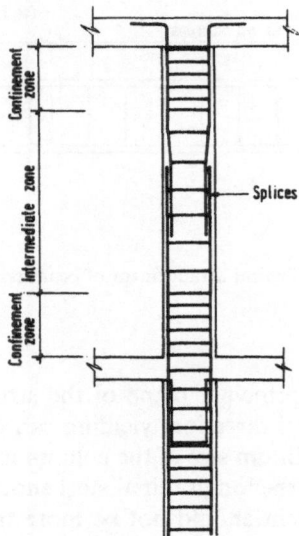

Fig. 6.25. Typical arrangements of column reinforcements.

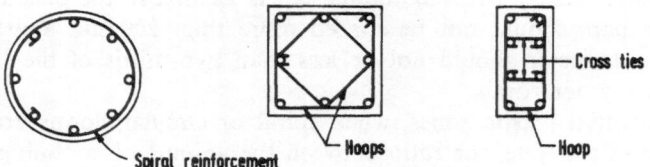

Fig. 6.26. Recommended arrangement of transverse reinforcement in columns.

Beam–Column Joint

The principal modes of failure of beam–column joints are

(a) shear failure

(b) anchorage failure of beam bars anchored in the joint

(c) bond failure of beam or column bars passing through the joint

For the joint to be strong enough to withstand the yielding of connecting beams or columns (occasionally), it is recommended that the confinement reinforcement should continue into the core of the joint and there should be sufficient reinforcement to carry the ultimate shear transmitted by the beams.

The shear strength of the joint is determined approximately using the shear panel analogy. From Fig. 6.26, the horizontal shear in the connection is given by

$$V = (A_{s1} + A_{s2})f_y^* - V_c \tag{6.18}$$

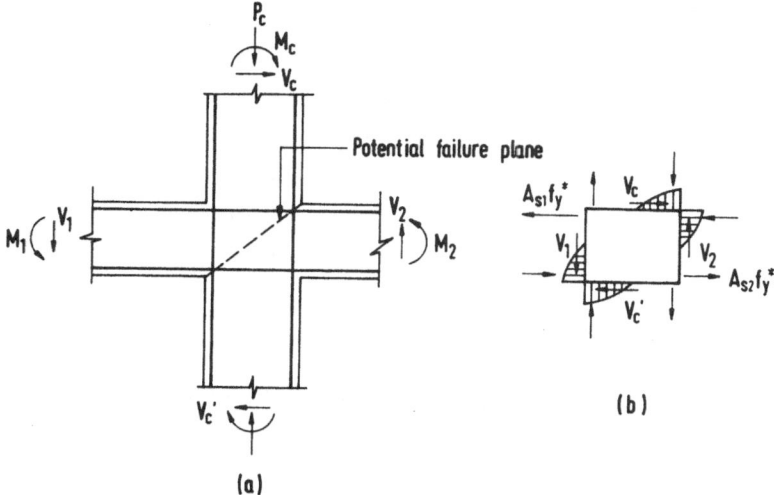

Fig. 6.27. Forces in a beam–column joint: (a) equilibrium of forces around the joint, (b) internal steel and concrete forces at the joint.

where V_c is the shear in the column above the joint, f_y^* is the factored yield strength which allows for over-strength, and A_{s1}, A_{s2} are the areas of tensile reinforcement.

The shear is resisted by the compressive strut action of concrete and tension in the horizontal stirrups, which must cross the failure plane indicated in Fig. 6.27. The areas of stirrups may be calculated conservatively as

$$A_{sh} = V/f_y \tag{6.19}$$

In order to avoid compression strut failure, the shear stresses in the joint need to be limited to $240A_j\sqrt{f_c'}$ (N/mm²), where A_j is the cross-sectional area of the joint and f_c' the compressive strength of concrete.

The anchorage failure of the beam bars in the external joint is avoided by providing either sufficient anchorage length within the joint or beam stubs, so that the longitudinal reinforcements may be anchored outside the core of the joint. Bond failure of beam or column bars is avoided by limiting the diameter of the bars.

6.4.2 Shear Wall Structures

Shear walls are excellent elements for seismic-resistant structures provided they are properly designed and detailed for strength and ductility. When shear walls are used together with rigid frames, the rigidity of the wall reduces the deflection demands on the other parts of the structure, such as column–beam joints.

Figure 6.28 depicts the principal modes of failure of a shear wall. The possibility of failure in any of these modes reduces with increase in axial load;

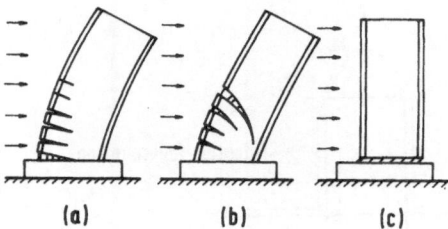

Fig. 6.28. Failure modes of cantilever shear wall: (a) flexural failure, (b) shear failure, (c) sliding failure.

however if the axial load is increased beyond a certain limit, the accompanied reduction in ductility may offset the increase in strength.

When the height of the wall is greater than twice the depth of the wall, the bending strength is calculated in the same way as the column provided the reinforcements are concentrated near the extreme fibres. It is recommended that much of the flexural steel of the wall be placed at the extreme fibres with a minimum of 0.25% of the vertical bars in the remainder of the wall, in order to enhance the curvature ductility of the wall.

The hinge zone at the base of the wall is taken as the length on plan or one-sixth of the total height, but not more than twice the length on plan [6.18]. In the hinge zone, the main reinforcements need to be restrained in the same manner as the columns. In the ultimate failure state (Fig. 6.28a), the tension steel yields, resulting in large ductility and energy dissipation.

A shear wall with a small aspect ratio could fail in shear (Fig. 6.28b) with diagonal cracks. If the horizontal reinforcement content is small, diagonal tension failure occurs; whereas if the vertical reinforcement is inadequate, diagonal compression failure occurs. Thus adequate shear strength for the wall is provided by following the rules for columns.

In the sliding shear failure (Fig. 6.28c), the shear wall moves horizontally. Vertical reinforcements uniformly spaced in the wall and diagonal reinforcements are effective in preventing this mode of failure.

6.4.3 Coupled Shear Walls

In the case of coupled shear walls (Fig. 6.29), failure will be concentrated around the openings and at the base. The coupling beams are subjected to a high ductility demand and they could fail in diagonal tension if they are too stiff.

The coupled walls are designed in such a way that plastic hinges first occur in the coupling beams, followed by hinges at the base of each wall. When the depth of the coupling beam is greater than the clear span, it is difficult to achieve sufficient ductility with the normal reinforcement arrangement. In such cases the recommended arrangement is given by Park and Paulay [6.19] and is shown

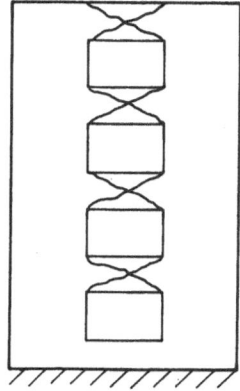

Fig. 6.29. Shear failure of walls with openings.

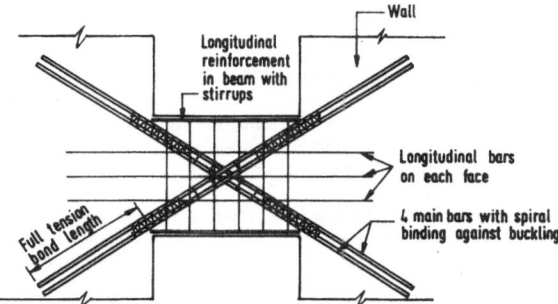

Fig. 6.30. Typical reinforcement arrangement for deep coupling beams.

in Fig. 6.30. The area of steel in each arm of the 'X' configuration is given by

$$A_s = \frac{V_u}{2f_y \sin \alpha} \tag{6.20}$$

where α is the inclination of each arm and V_u the ultimate shear. The ultimate moment of resistance is then given by

$$M_u = V_u \frac{l_s}{2} \tag{6.21}$$

where l_s is the clear span of the coupled beam. The above reinforcement is found to give far superior ductility than the conventional one.

Many papers on the seismic design of concrete structures are available. For example, seismic design of concrete tubular systems is given by Iyengar and Iqbal [6.20]. Fintel and Ghosh [6.21] have discussed what modifications with regard to strength and inelastic deformability would be required when a building proportioned to resist wind load is subjected to varying levels of earthquake intensities.

References

6.1. Walpole W R and Butcher G W, Beam design, seismic design of steel structures study group, section C, *Bulletin of the New Zealand National Society for Earthquake Engineering* 1985; 18: 337–343.
6.2. Butterworth J W and Spring K C F, Column design, seismic design of steel structures study group, section D, *Bulletin of the New Zealand National Society for Earthquake Engineering* 1985; 18: 345–350.
6.3. Wilteveen J H, Stark J W B, Bijlaard F S K and Zoetemeizer P, Welded and bolted beam to column connections, *Journal of Structural Division, ASCE* 1982; 108: 433–455.
6.4. Walpole W R, Beam–column joints, seismic design of steel structures study group, section H, *Bulletin of the New Zealand National Society for Earthquake Engineering* 1985; 18: 369–380.
6.5. Walpole W R, Concentrically braced frames, seismic design of steel structures study group, section E, *Bulletin of the New Zealand National Society for Earthquake Engineering* 1985; 18: 351–354.
6.6. Hjelmstad K D and Popov E P, *Seismic Behaviour of Active Beam Links in Eccentrically Braced Frames*, Report No. UBC/EERC–83/15, Earthquake Engineering Research Center, University of California, Berkeley, 1983.
6.7. Kasai K and Popov E P, General behaviour of WF steel shear link beams, *Journal of Structural Division, ASCE* 1986; 112: 362–382.
6.8. Balendra T, Lam K Y, Liaw C Y and Lee S L, Behaviour of eccentrically braced frame by pseudo-dynamic test, *Journal of Structural Division, ASCE* 1987; 113: 673–688.
6.9. Whittaker A S, Uang C and Bertero V V, *Earthquake Simulation Tests and Associated Studies of a .3 Scale Model of a Six-storey Eccentrically Braced Steel Structure*, Report No. UBC/EERC–87/02, University of California, Berkeley, 1987.
6.10. Nishiyama I, Midorikawa M and Yamanouchi H, Inelastic behaviour of full scale eccentrically K-braced steel building, *Proceedings of the 9th World Conference on Earthquake Engineering* 1988; 4: 261–266.
6.11. Balendra T, Sam M T and Liaw C Y, Diagonal brace with ductile knee anchor for aseismic steel frame, *Earthquake Engineering and Structural Dynamics* 1990; 19: 847–858.
6.12. Balendra T, Sam M T, Liaw C Y and Lee S L, Preliminary studies into the behaviour of knee braced frames subjected to seismic loading, *Journal of Engineering Structures* 1991; 13: 67–74.
6.13. Sam M T, A new Knee-Brace-Frame system for seismic resistant steel buildings, PhD thesis, National University of Singapore, 1992.
6.14. Libby J R, Eccentrically braced frame construction – a case history, *Engineering Journal, American Institute of Steel Construction* 1981; 4: 149–153.
6.15. Wang C, Structural system – Getty Plaza tower, *Engineering Journal, American Institute of Steel Construction* 1984; 4: 49–59.
6.16. Uniform Building Code, *International Conference of Building Officials*, 1978 edn, Whittier, California, 1978.
6.17. ACI Committee 318, *Building Code Requirements for Reinforced Concrete* (ACI 318–83), American Concrete Institute, Detroit, 1983.
6.18. *New Zealand Standard Code of Practice for the Design of Concrete Structures* (NZS 3101: 1982), Standard Association of New Zealand, 1982.
6.19. Park R and Paulay T, *Reinforced Concrete Structures*, Wiley, New York, 1975.
6.20. Iyengar H and Iqbal M, Seismic design of composite tubular buildings. In: Beedle L S (editor), *Advances in Tall Buildings*, Council of Tall Buildings and Urban Habitat, Van Nostrand Reinhold, New York, 1986, pp 133–148.
6.21. Fintel M and Ghosh S K, Earthquake resistance of buildings designed for wind. In: Beedle L S (editor), *Advances in Tall Buildings*, Council of Tall Buildings and Urban Habitat, Van Nostrand Reinhold, New York, 1986, pp 461–472.

Subject Index